WORLD BOOK

AF862747

SHAPES AND NUMBERS

CHILDCRAFT

www.worldbook.com

Co-published by agreement between Shi Tu Hui and World Book, Inc.

Shi Tu Hui
Room 1807, Block 1,
#3 West Dawang Road
Chaoyang District, Beijing 100025
P.R. China

World Book, Inc.
180 North LaSalle Street
Suite 900
Chicago, Illinois 60601
USA

Library of Congress Control Number: 2024943586

Childcraft: The How & Why Library
ISBN: 978-0-7166-5548-0 (set, hard cover)

Shapes and Numbers
ISBN: 978-0-7166-5555-8 (hard cover)
ISBN: 978-0-7166-5565-7 (soft cover)
ISBN: 978-0-7166-5575-6 (e-book)

WORLD BOOK STAFF

Editorial

Vice President
Tom Evans

Editorial Project Coordinator
Nathalie Strassheim

Manager
Nicholas Kilzer

Senior Content Creator
Jenna Neely

Senior Editors
Emma Flickinger
William M. Harrod

Curriculum Designer
Caroline Davidson

Indexing
Nathaniel Lindstrom

Freelance Writer
Alex Woolf

Graphics and Design

Senior Visual Communications Designer
Melanie Bender

Senior Designer
Isaiah Sheppard

Digital Asset Specialist
Rosalia Bledsoe

Freelance Designer
Starletta Polster

Contents

Introduction

Our world is full of shapes. Just look around you, and you will see them everywhere. The pages of this book are shaped like rectangles. When you eat lunch, you may have a sandwich on square bread. If the bread is cut from corner to corner, you have two triangles. A cookie for dessert may be shaped like a circle.

Our world is full of numbers, too. You can see them on calendars, cell phones, and price tags. And think of all the numbers that tell something about you. Your age, your grade in school, and your birthday are just a few.

This book, *Shapes and Numbers*, will take you on a treasure hunt for those everyday shapes and numbers. You will look at the shapes in farm fields, tile floors, and Ferris wheels. You'll find patterns in a flower garden and puzzles in a box of buttons.

You'll figure out what your chances of correctly guessing outcomes are when you flip a coin or roll dice. You'll discover how to measure your height, tell time, and read a thermometer.

The book will also lead you on a journey through the mysteries of mathematics. You will go back in history to discover how and why people began to count. You'll learn to write numbers the way people did thousands of years ago, and you'll find out where our numbers came from. You'll uncover the secrets of magic squares and other number tricks. In the process, you'll also see how useful—and fun—it is to know the arithmetic basics of adding, subtracting, multiplying, and dividing.

There are many features in this book to help you find your way through it. You will find fun-filled facts in the boxes marked **KNOW IT ALL**. You can amaze your friends with what you learn!

This book also has many activities that you can do at home. Look for the words **Try This…** . The activity that follows offers a way to learn more about shapes and numbers. For example, you can make patterned jewelry to wear, or you can use numbers to create special codes for writing and reading secret messages!

Some of the activities are very easy. Others may require a little adult help with such tasks as cutting or measuring. And then there are some really fun ones, but they are a bit more complex and require more materials. Take a moment to review the list of materials needed and to read through the step-by-step instructions before you begin.

As you read this book, you will see that some words are printed in bold-face type, **like this**. These are words that might be new to you. You can find the meanings and pronunciations of these words in the Glossary at the back of the book. Turn to the Index to look up page numbers of subjects that particularly interest you.

Fun with Shapes

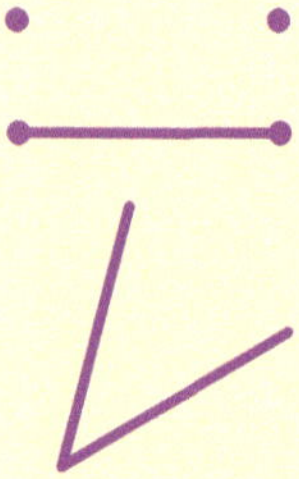

Start with two points.
Connect them with a line.
Two lines can make an angle —
that looks fine!

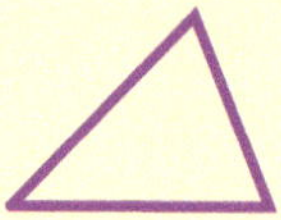

Add another line
and a triangle's there.
Or use four lines to make a rectangle or square.

Angles, squares, triangles,
rectangles, points, and lines—
we can use them all to make shapes and designs!

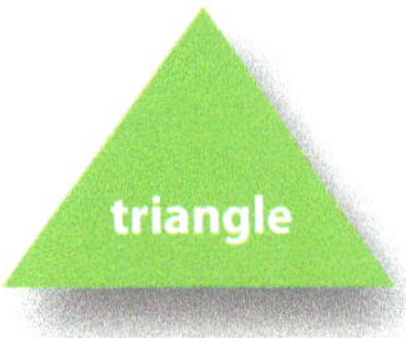

Shapes Everywhere

Think of all the shapes in your world! Your toys and games, for example, are made up of many shapes. A checkerboard is a square on which other squares are printed. The top of a checker is a circle. A sail on a toy boat is a triangle. Each card in a deck is a rectangle.

Look for shapes in other things you use. A book cover is a rectangle. A DVD is a circle. What other shapes can you find?

Rectangles, squares, triangles, and circles are flat. Some other shapes have thickness. A pointed ice-cream cone is named for its shape—the **cone**. Sugar cubes are also named for their shape, the **cube**.

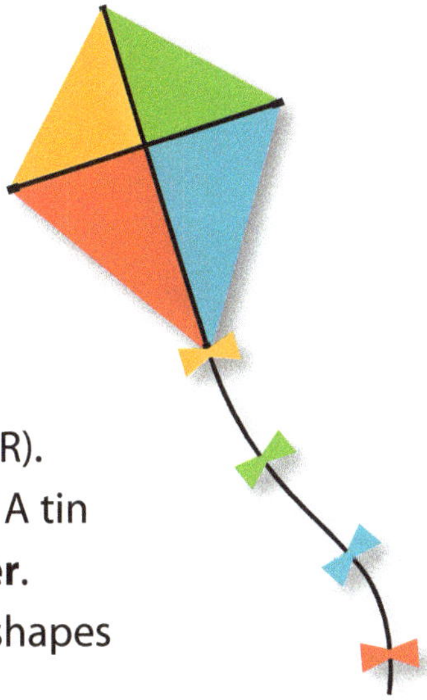

A basketball is a **sphere** (SFIHR). Can you think of other spheres? A tin can has a shape called a **cylinder**.

Look at these pictures. What shapes can you find?

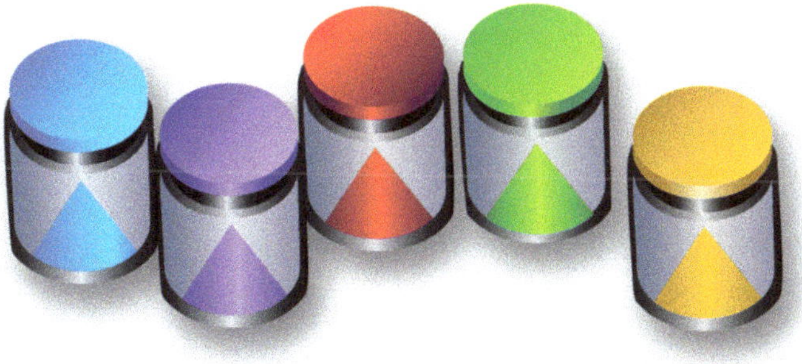

Lines and Points

What is between this sentence and the next one?

You probably said, "It's a line." Lines are everywhere. People use lines to mark things like boundaries on a sports field, highway lanes, and parking spaces.

But what is a line? It's a long, thin mark. It can be straight or curved. A straight line is the shortest distance between two points.

A point has no **length** or width. It only has a position. To show its position, we can draw a dot:

Try This...

You can use dots and lines to make a picture like the one at the bottom of this page. All you need are paper and a pencil or marker. Think of a shape—for example, a rectangle. Draw four dots for the corners. Then connect them with straight lines.

Now turn your rectangle into part of a bigger picture, like a house. Draw more dots to show the outline of the picture. Then connect them with straight lines.

You can even make a dot picture for a friend to finish! Draw dots for the outline. Number the dots to show which ones should be connected. Let your friend draw lines from dot to dot, starting with 1.

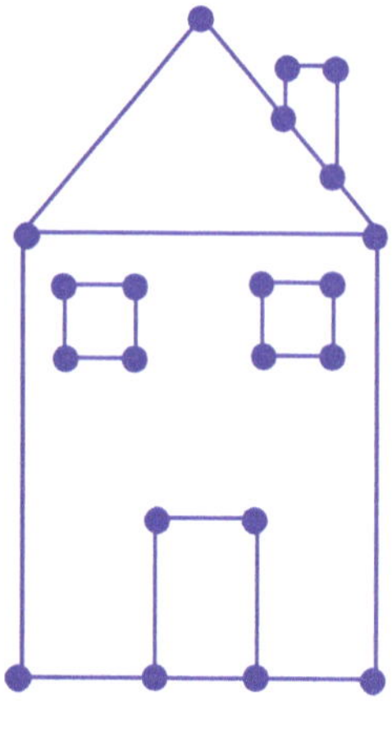

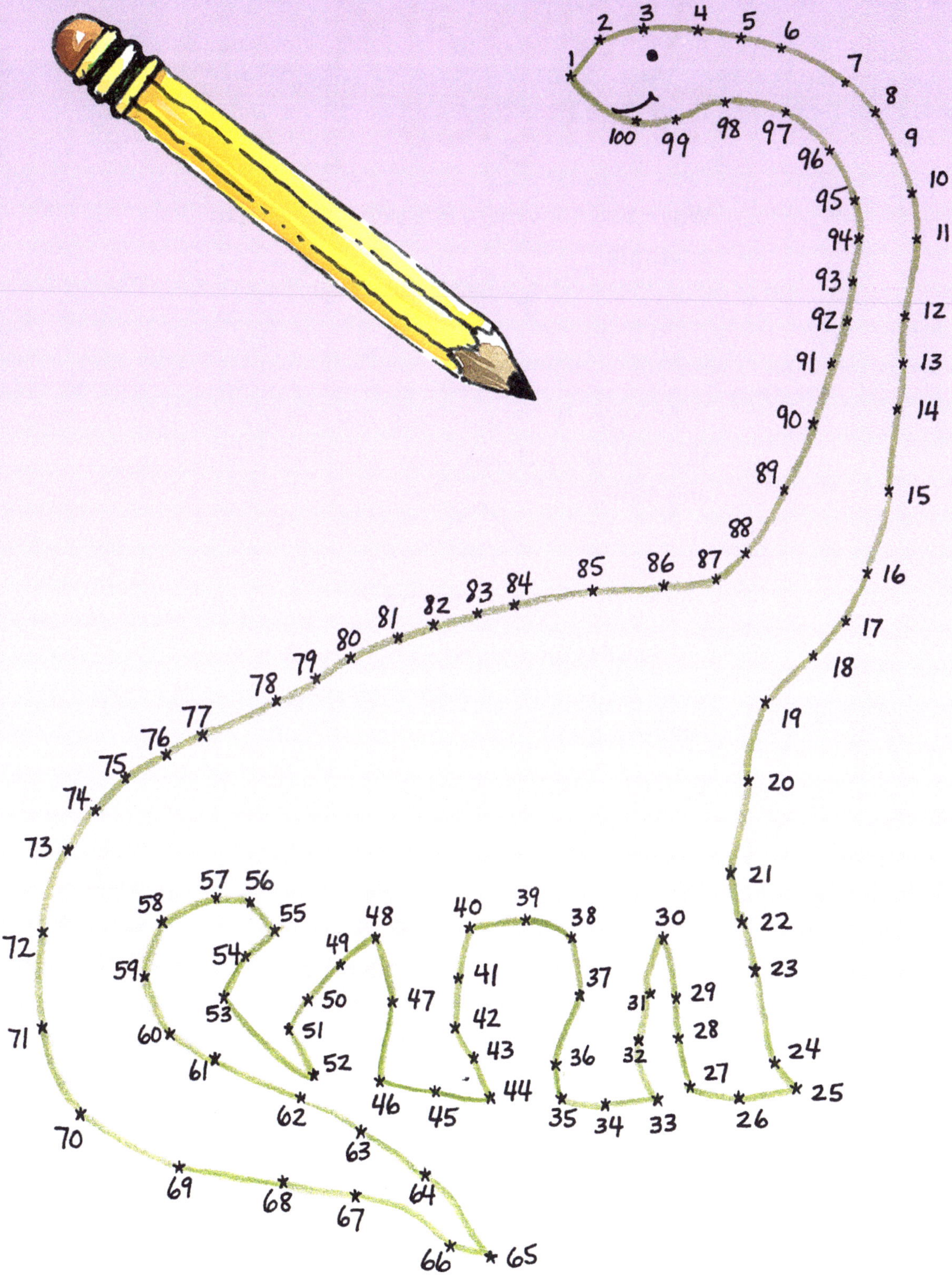
1
2
3
4
5
6
7
8
9
10
11
12
13
14
15
16
17
18
19
20
21
22
23
24
25
26
27
28
29
30
31
32
33
34
35
36
37
38
39
40
41
42
43
44
45
46
47
48
49
50
51
52
53
54
55
56
57
58
59
60
61
62
63
64
65
66
67
68
69
70
71
72
73
74
75
76
77
78
79
80
81
82
83
84
85
86
87
88
89
90
91
92
93
94
95
96
97
98
99
100

Angles and Triangles

When two lines meet at a point, they make an **angle.**

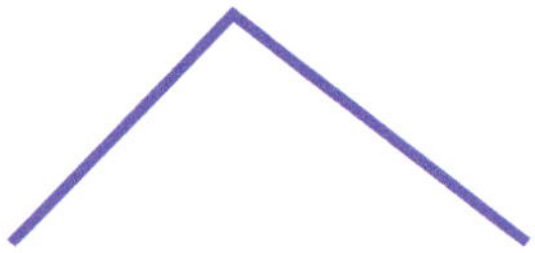

An angle has two sides. If you add a third side, you get a shape called a triangle (TRY ang guhl).

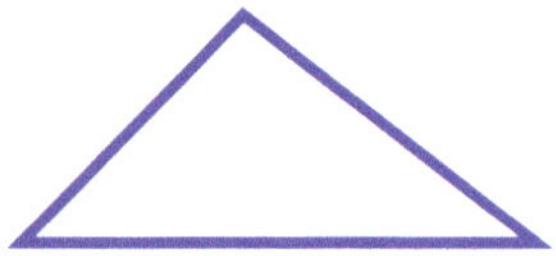

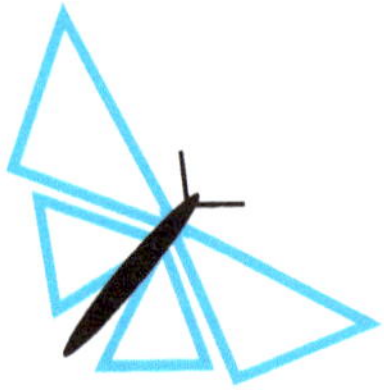

Triangles have three straight sides and three angles. The word *triangle* means "three angles."

An equilateral (EE kwuh LAT uhr uhl) triangle is a special kind of triangle. Its sides are all the same length. Its angles are all the same size. The word *equilateral* means "equal sided."

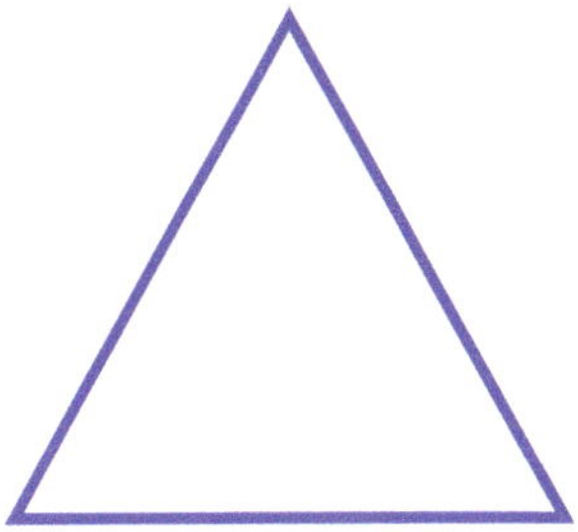

Try This...

Make a picture. Start with angles and triangles. Then add details to finish your picture. What did you make?

Try This...

Counting Triangles

The design on this page is made of 88 triangles. Can you find them all? Cover the next page while you look!

Here's one more question for you: What kind of triangles are they?

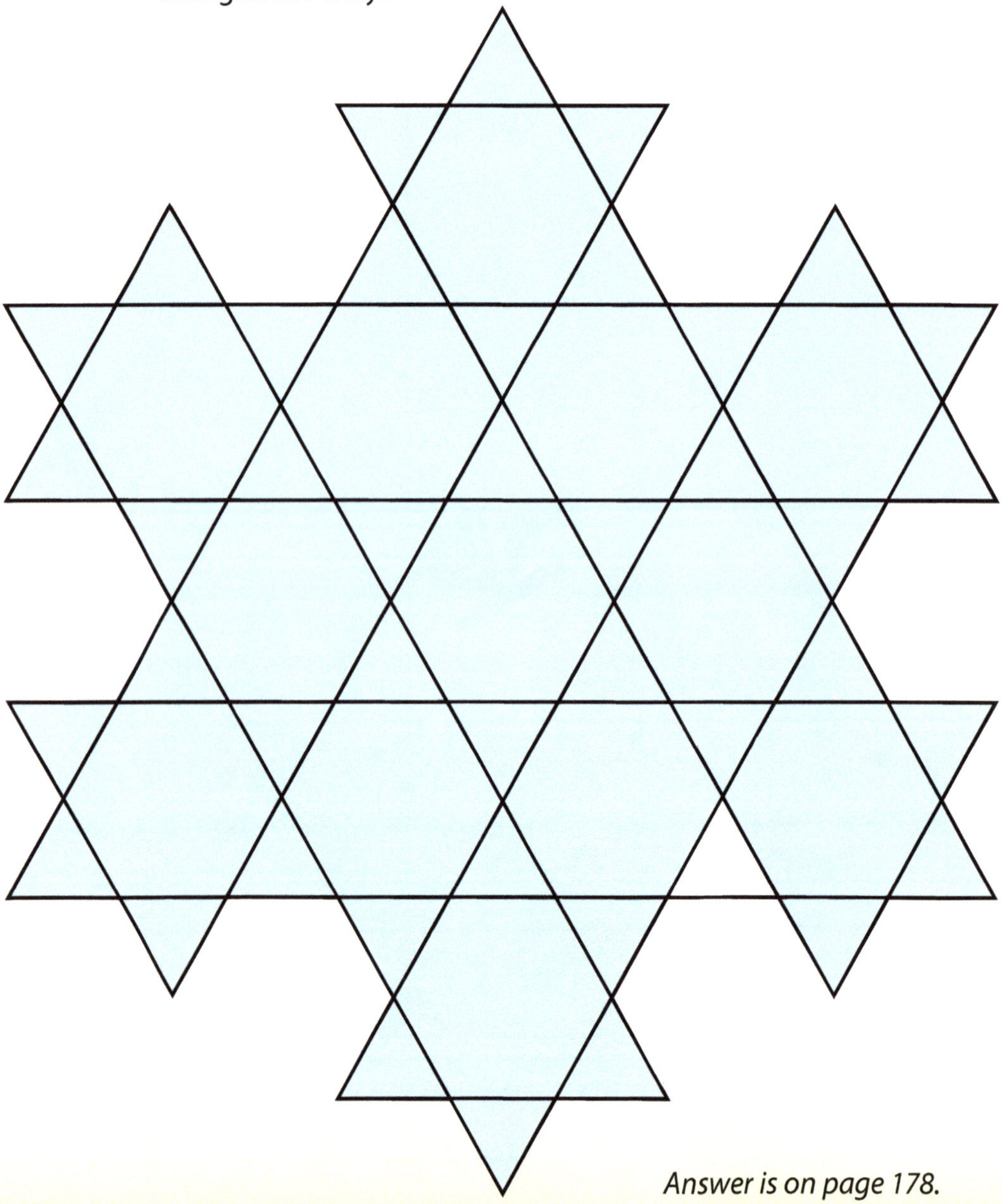

Answer is on page 178.

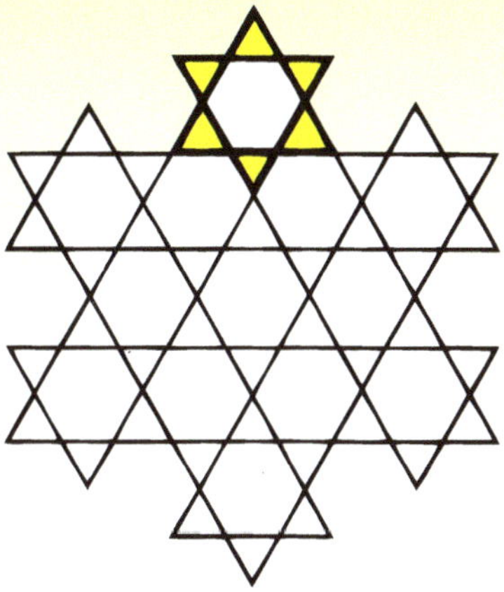

There are 42 triangles of this size. They point in all directions.

There are 26 triangles of this size, 13 pointing up and 13 pointing down.

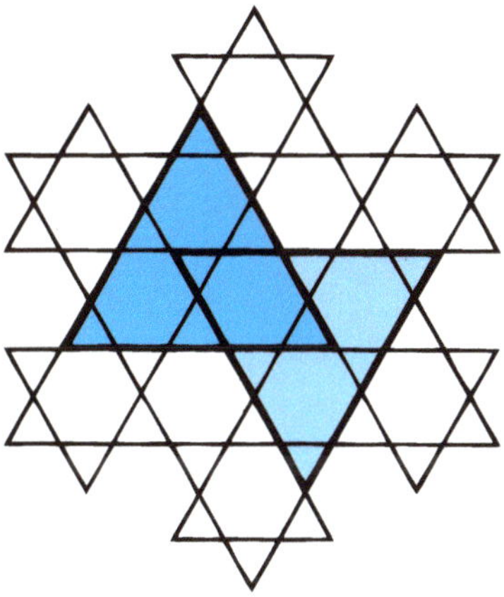

There are 12 triangles of this size, 6 pointing up and 6 pointing down.

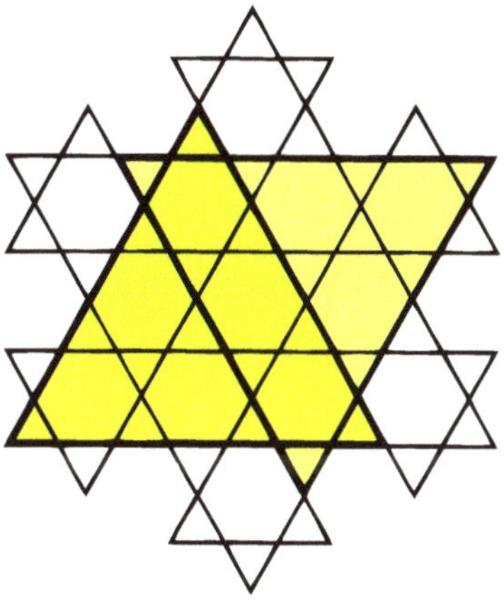

There are 6 triangles of this size, 3 pointing up and 3 pointing down.

Finally, there are 2 triangles of this size, 1 pointing up and 1 pointing down.

That makes a total of 88 triangles. Did you find them all?

Try This...

Disappearing Squares

A square is a shape with four sides. All of its sides are the same length, and all of its angles are the same size.

Use four toothpicks, craft sticks, or straws to make a square. See the four angles? Angles of this size are called right angles.

Now try these puzzles:

You Will Need:

- 17 toothpicks, craft sticks, or straws

What To Do:

1. Put 17 sticks together to make six squares, like this:

   Now take away four sticks so that only four squares are left.

   There's more than one way to solve this puzzle. Can you find at least two solutions?

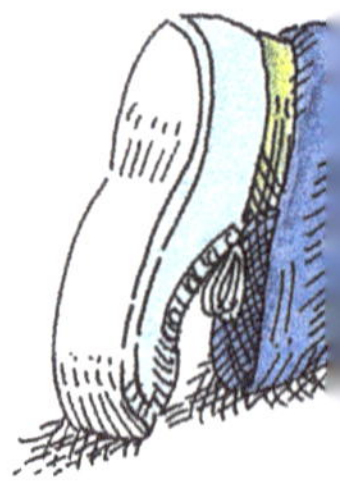

2. Put your six squares back together. Now take away four sticks so that only three squares are left. *(Hint: All the squares don't have to be the same size.)*

3. Put 12 sticks together to make a "tic-tac-toe" board like this:

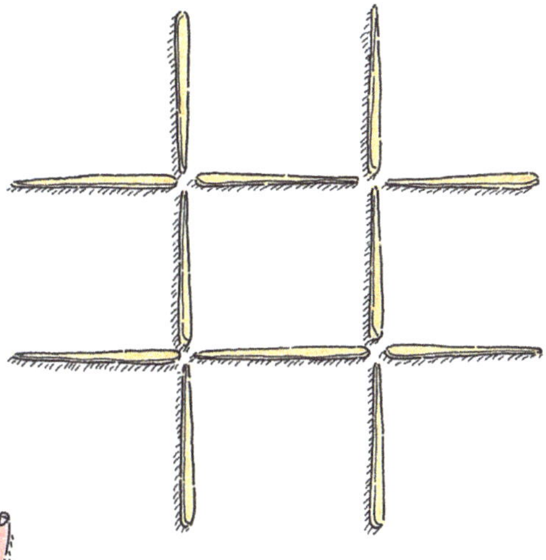

Now see if you can:

a. Make four squares by moving four sticks.

b. Make three squares by moving four sticks.

c. Make two squares by moving six sticks.

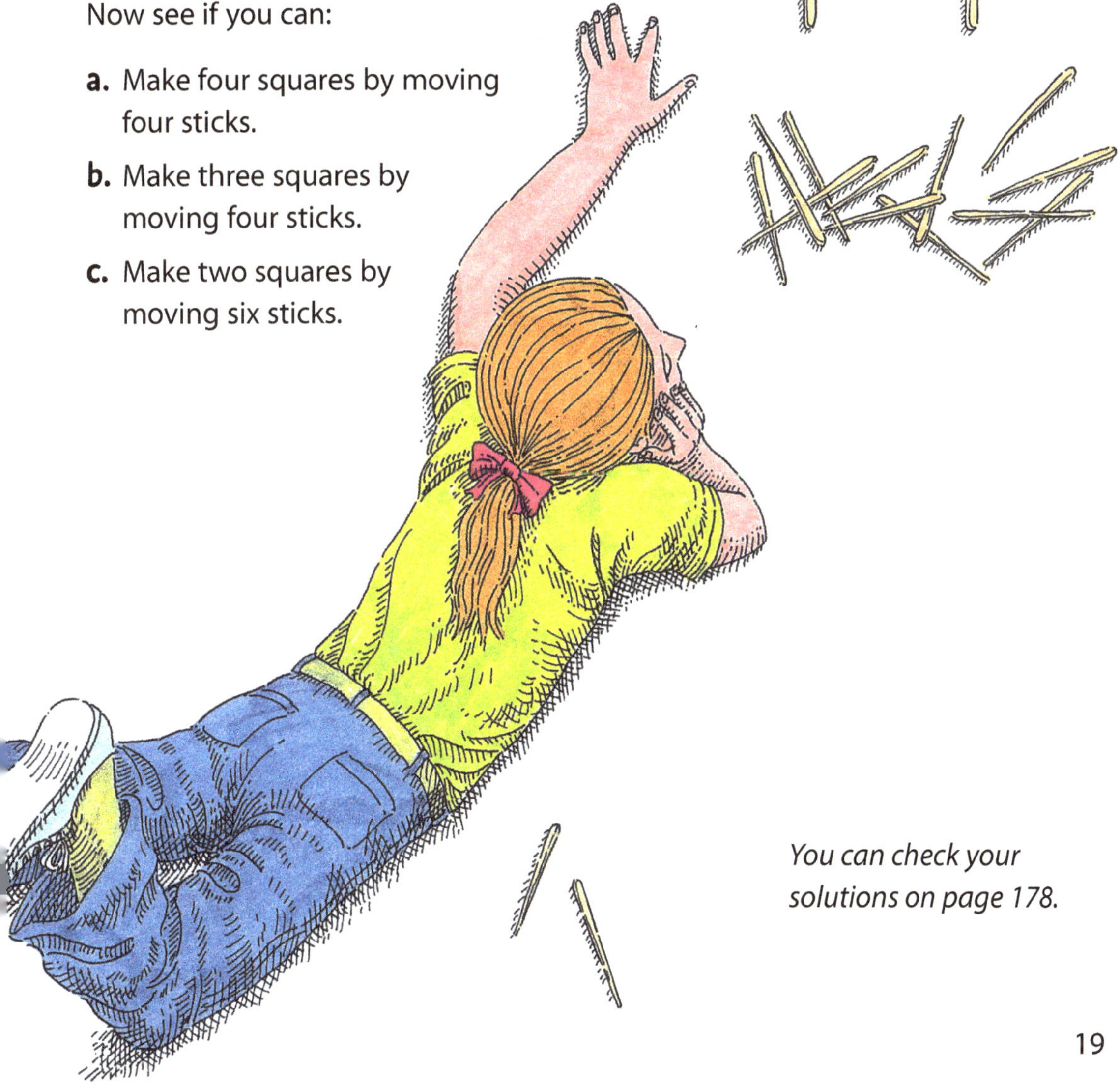

You can check your solutions on page 178.

The Farmer's Square

Long ago, there was a farmer whose land was in the shape of a square. Each side of the square was exactly 100 paces long.

One day, a tired, dusty man knocked at the farmer's door and asked for something to eat. The kind farmer gave him a nice lunch.

After the man had eaten, he said, "Farmer, I am your king! As a reward for your kindness, you may double the size of your land. But after you add the new land, your farm must still be in the shape of a square."

The farmer was overjoyed. He went out at once to **measure** his new land and put a fence around it.

At first the farmer thought the sides of the new square should be 200 paces long—twice as long as the sides of the old square. But this plan didn't work. The square he measured was four times as big as his old farm!

The farmer thought for a long time. Then he got an idea. He would divide his land into four smaller squares. Then he could add on four new squares of land the same size as the small squares. That way his new farm would be exactly twice as big as his old farm.

The farmer drew his plan on a piece of paper. First he added four small squares onto one side of his old square. But this gave him a rectangle, not a square.

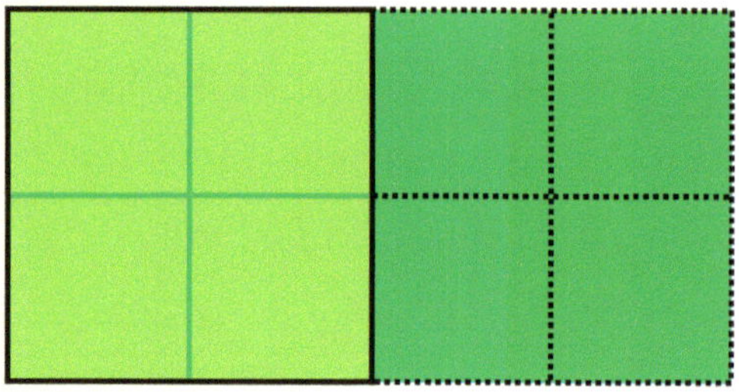

Then he added two small squares to one side of the old square and two to another side. This almost worked, but there was a piece missing out of one corner, so it wasn't a perfect square.

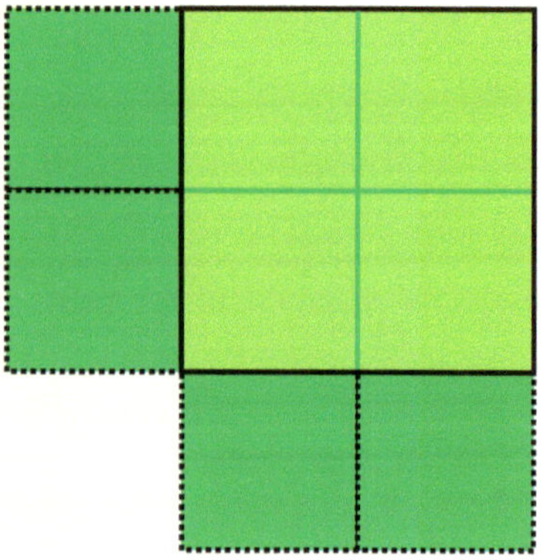

Then the farmer had another idea. He divided his old square into four triangles by drawing lines between the opposite corners. Could he add four more triangles of exactly the same size to his farm—and somehow make a square?

The farmer added a triangle to one side of the old square. Then he added a triangle to each of the other sides.

The farmer looked at his paper closely. Then he turned it sideways. What do you think he saw?

A square!

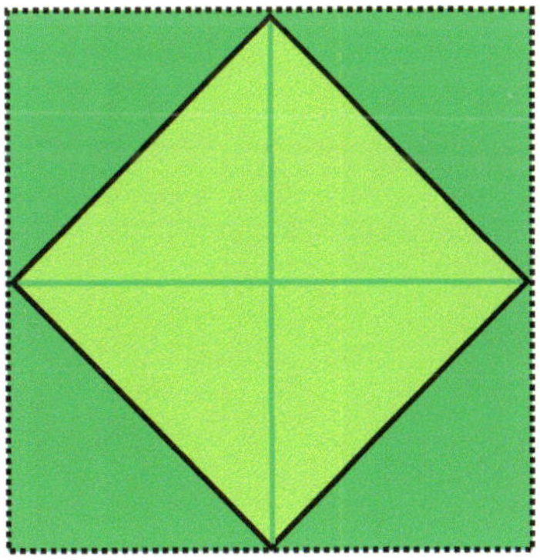

Now you know a secret: *You can use triangles to double the size of a square—and still have a square!*

Polyominoes

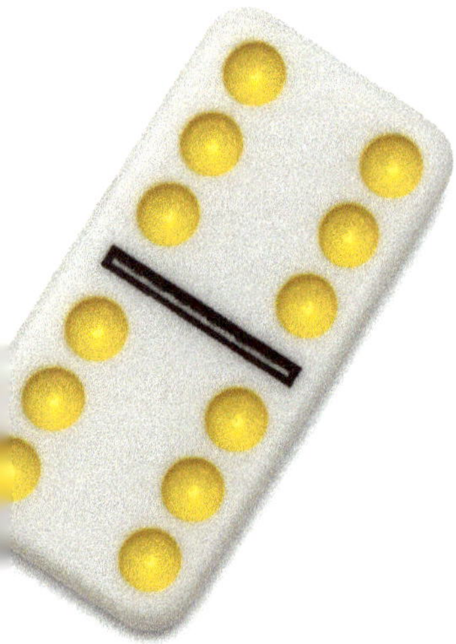

A polyomino (pahl ee AHM uh noh) is a shape made from squares. The squares must be joined by at least one full side.

Polyominoes are named for the number of squares they contain. A domino is a shape made from two squares. Have you ever played a game of dominoes?

Shapes made from three squares are called trominoes. A four-square shape is a tetromino. How many different tetrominoes can you draw?

Shapes made from five squares are called pentominoes (pehn TAHM uh nohz). There are 12 different pentominoes.

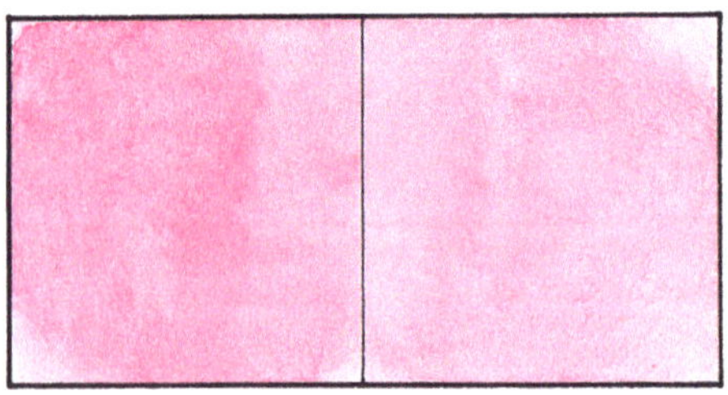

domino

tromino

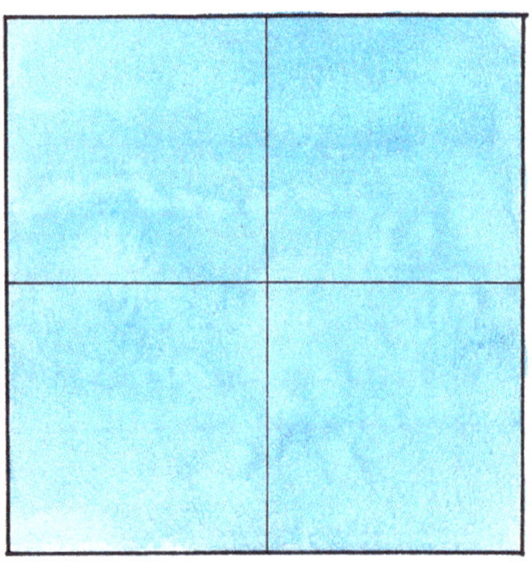

tetromino

Try This...

Draw the 12 pentominoes below on graph paper (paper with tiny squares all over it). Cut them out. Then fit the pieces together to make a rectangle. You can turn or flip the pieces as needed.

First try making a 6 x 10 rectangle. That's a rectangle with 6 squares on two sides and 10 squares on the other two sides. Then try making a 5 x 12 rectangle.

There are more than a thousand ways to make each rectangle! You will find two solutions on page 179.

pentominoes

Around and Around

Did you ever ride a Ferris wheel or a carousel? These rides—and many other rides at fairs and amusement parks—are made with circles.

A circle is a closed curve. Every point on the circle is the same distance from the center. This distance is called the radius (RAY dee uhs).

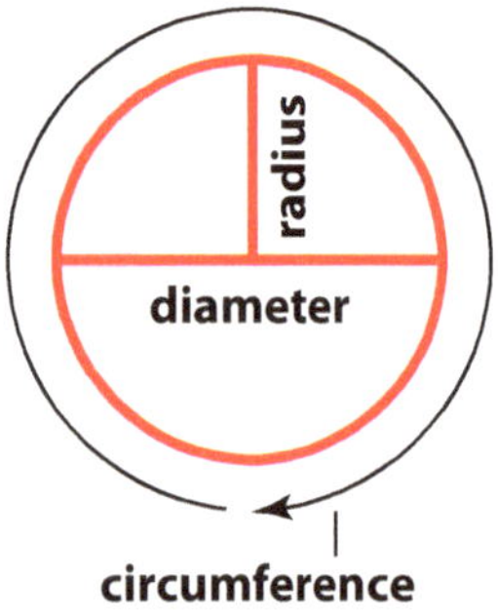

The distance across a circle, going through the center, is called the diameter (dy AM uh tuhr). The diameter is always twice as long as the radius.

The distance around the edge of a circle is called the circumference (suhr KUHM fuhr uhns).

Try This...

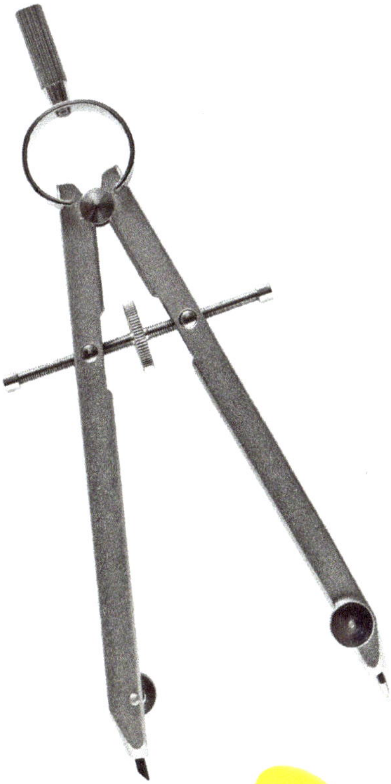

You can easily draw a circle with a tool called a **compass**. If you don't have a compass, try drawing a circle this way:

1. Lay a piece of paper on thick cardboard. Press a pushpin into the center of the paper.

2. Tie one end of a short piece of string, about 4 inches (10 centimeters) long, to the pushpin. Tie the other end of the string around a pencil. Make sure the string is tied tight enough not to fall off.

3. Stretch out the string and put the tip of the pencil on the paper. Hold the pencil upright. Move the pencil around the pushpin, keeping the string pulled tight. When you reach your starting point, you'll have a circle!

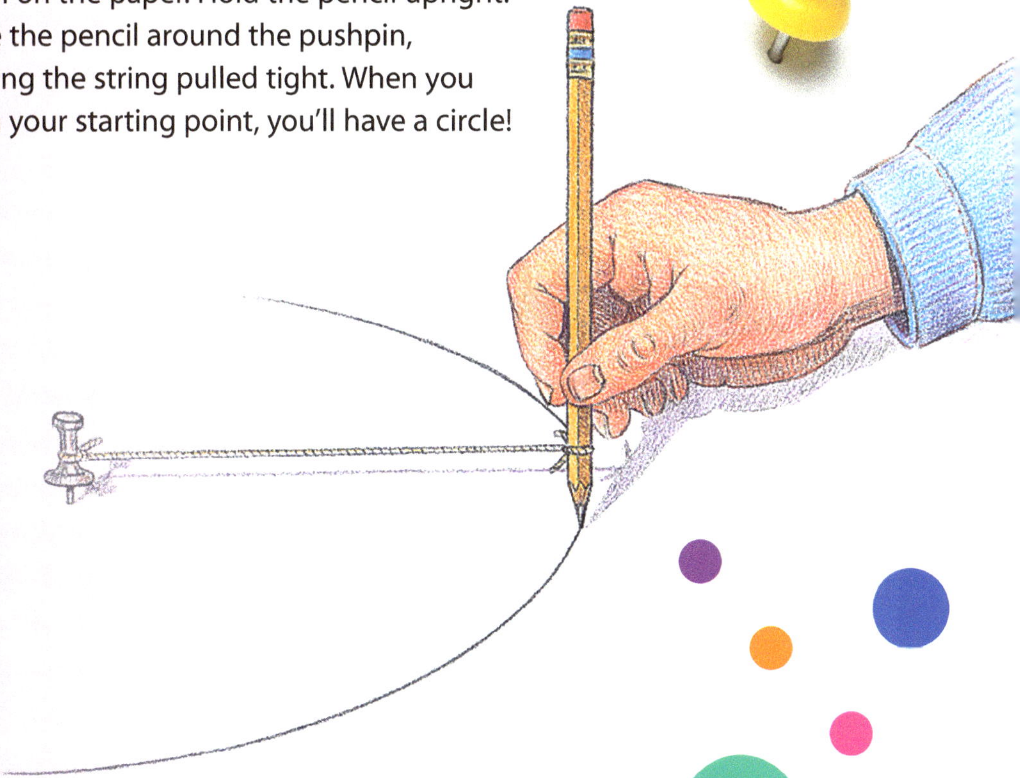

Circle of Stones

About 4,000 years ago, some people in what is now southwestern England made a circular monument of stones and ditches. How did they make their huge circles? They may have used pebbles, a leather rope, and a wooden stake.

Imagine this scene: One person pounded a stake into the ground and tied one end of a long leather rope around it. A second person tied the other end of the rope around his waist and walked away until the leather was stretched out tight. Then he walked slowly around the stake, keeping the leather tight. A third person followed, placing pebbles on the ground behind his friend's footsteps. The pebbles formed a circle.

Many people worked to dig a circular ditch. Later, other people placed huge stone blocks inside the circle. It took hundreds of years to build this monument. Today it is called *Stonehenge.*

Most scientists think Stonehenge was used as a giant calendar. Here, people kept track of time and dates by watching the movements of the sun and the moon over the circle of blocks. Many of the stones are missing or have fallen over, but scientists can still figure out what Stonehenge probably looked like long ago.

Stonehenge is an ancient monument in southwestern England. It was a circle of huge stones built between about 3100 and 1500 B.C.

Try This...

Shape Tangle

Here's a game with a special twist! You can play it with your friends.

You Will Need:

- 3 large triangles
- 3 large squares
- 3 large circles made of construction paper
- masking tape

What To Do:

1. Find two or more friends to play with you. The more players you have, the more fun the game will be!
2. Hold the shapes over your head and drop them.
3. Tape the shapes to the floor where they land. (Ask permission first!)
4. Pick one player to be the "caller." The rest of the players should stand around the shapes.
5. The caller gives directions, such as "Put your elbow on a square," or "Touch your right knee to a triangle."

6. The other players must follow the directions. The first one or two directions will be easy, but after that, players usually must bend and twist to reach several shapes at once. Two or more players can touch the same shapes.
7. Players who fall down or cannot follow a direction are out of the game. The winning player gets to be the caller for the next game!

Try This...

Tangrams

The **pattern** below is made of different shapes. This pattern is called a tangram (TANG gram). It is one of the oldest puzzles in the world. Tangrams were popular in China thousands of years ago.

You Will Need:

- a piece of paper
- scissors
- a pencil
- a ruler

What To Do:

1. Lay your paper over the square on page 32 and trace the lines. Use the ruler to help you draw straight lines.

2. Cut out the seven shapes: two large triangles, two small triangles, one medium triangle, one square, and one **parallelogram**, which looks like a slanted rectangle.

3. Fit all the pieces together to make a rectangle like this one.

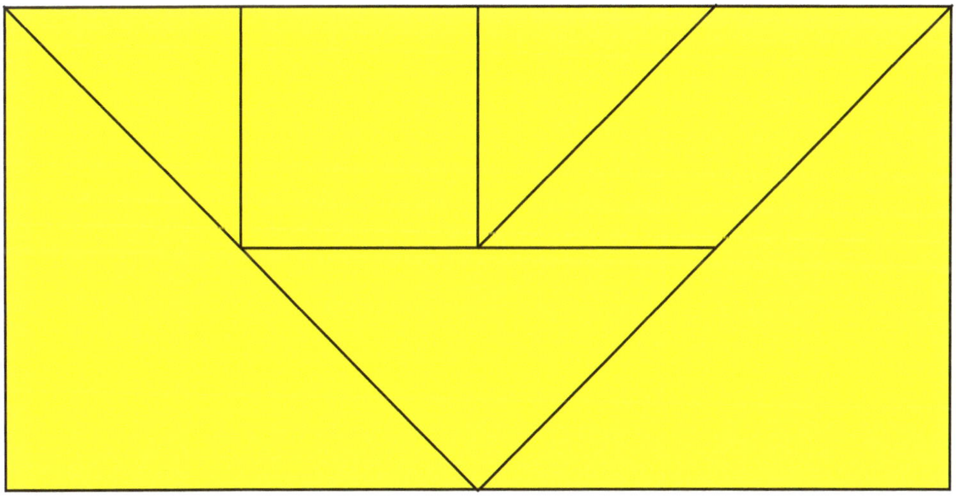

4. Now try to make these shapes with your tangram pieces—a dinosaur, a bird, and a boat.

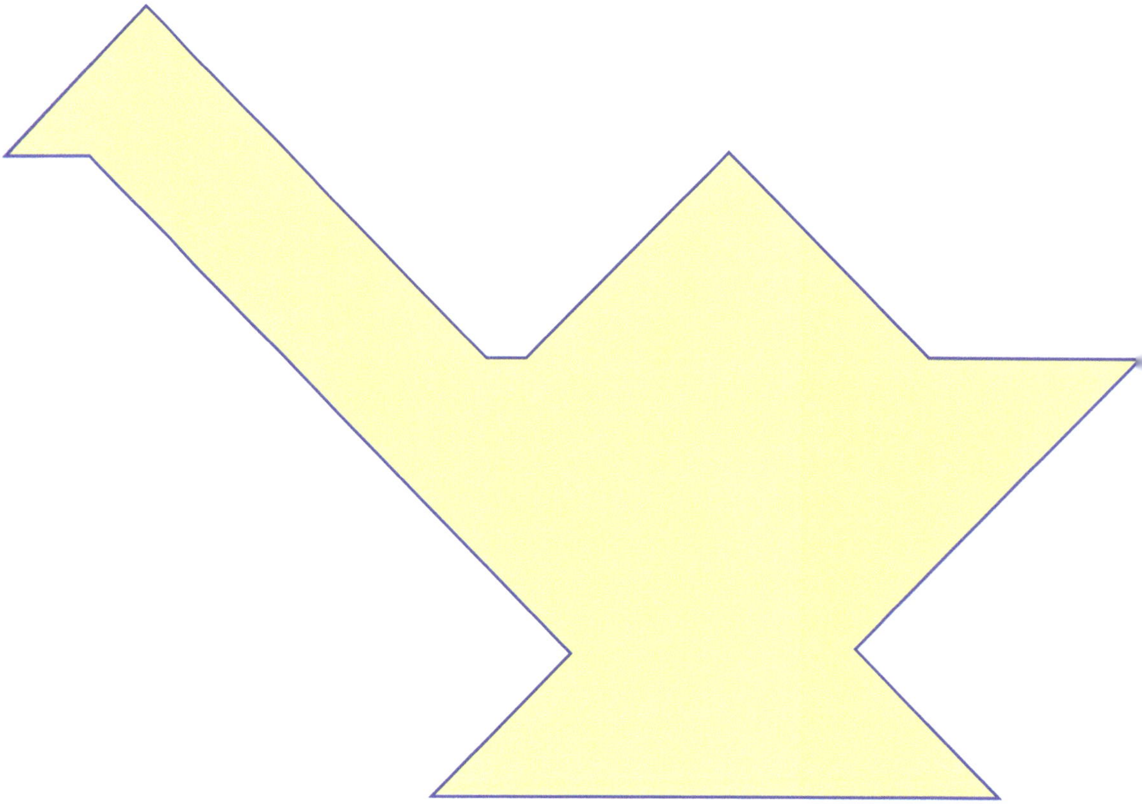

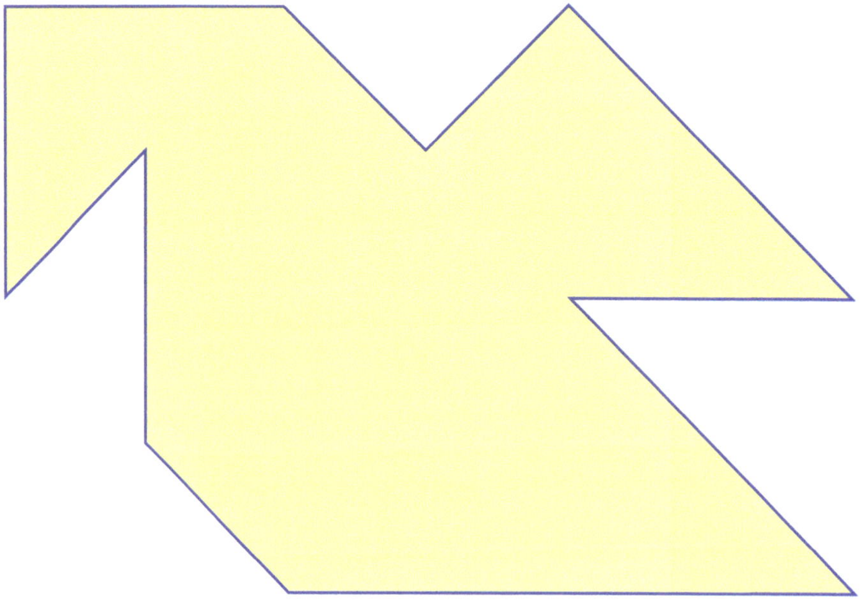

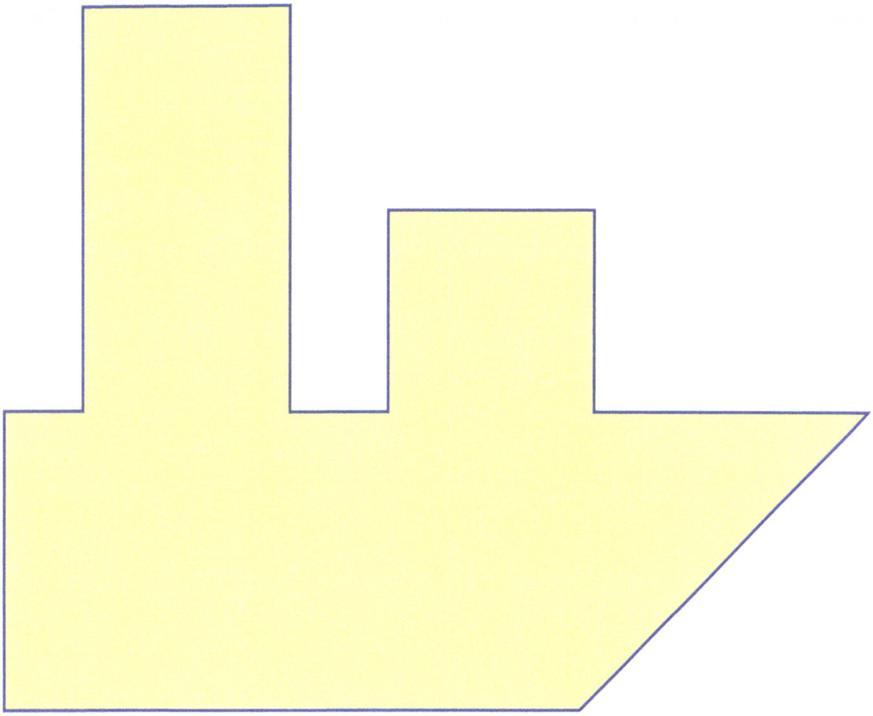

5. Make your own shapes with the tangram pieces. What else can you make?

Check your solutions on page 179.

Polygons

A **polygon** (PAHL ee gahn) is any flat shape whose sides are straight lines. We have already seen some of these shapes.

Polygons are named for the number of sides they have. Here are some more polygons:

A **pentagon** is a five-sided shape.

A **hexagon** is a six-sided shape.

How many sides does a **heptagon** have? That's right—seven!

An **octagon** has eight sides. It is a familiar shape in many countries because it is used for stop signs.

Here's a **nonagon**. Count the nine sides.

This is a 10-sided shape. It's called a **decagon**.

Try This...

Look carefully at the shapes on these pages. Could you fold each shape once to get two matching halves?

Take a guess. Then trace each shape on a piece of paper and cut it out. Try to fold each shape in half so that the halves match. Can you do it? Did you guess correctly?

Any shape you can fold to get two matching halves is called **symmetrical** (sih MEHT ruh kuhl).

Check your answers on page 179.

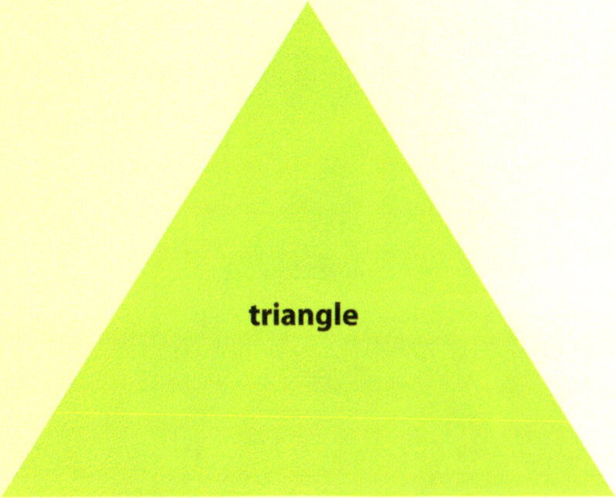

How many sides does a triangle have? *Tri* means "three."

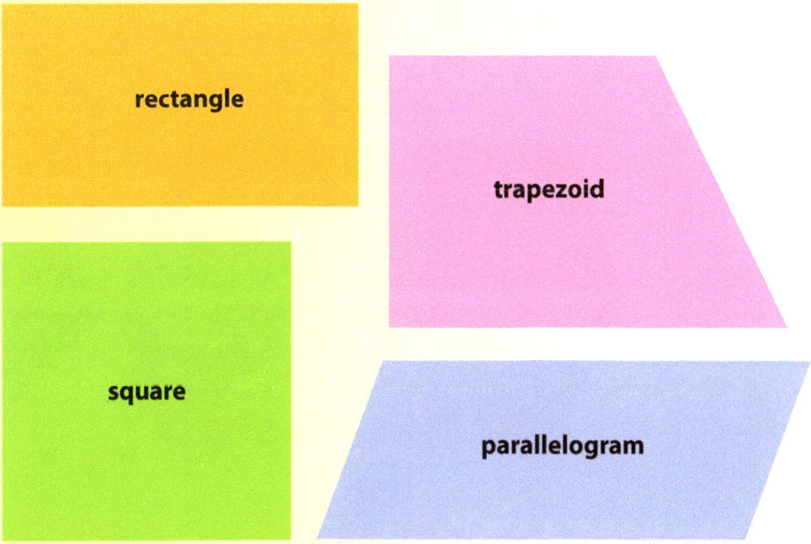

A **quadrilateral** has four sides. *(Quad* means "four.") Rectangles, squares, **trapezoids**, and parallelograms are all quadrilaterals.

Try This...

Building with Shapes

Look at these pictures. They're all made from shapes! You can make shape pictures, too.

You Will Need:

- a pencil
- white paper
- scissors
- colored construction paper
- glue (optional)

What To Do:

1. Trace the shapes on page 39 onto white paper.
2. Cut out the shapes.
3. Lay the shapes on colored construction paper. Hold each shape down and carefully draw around it.
4. Cut out the colored shapes.

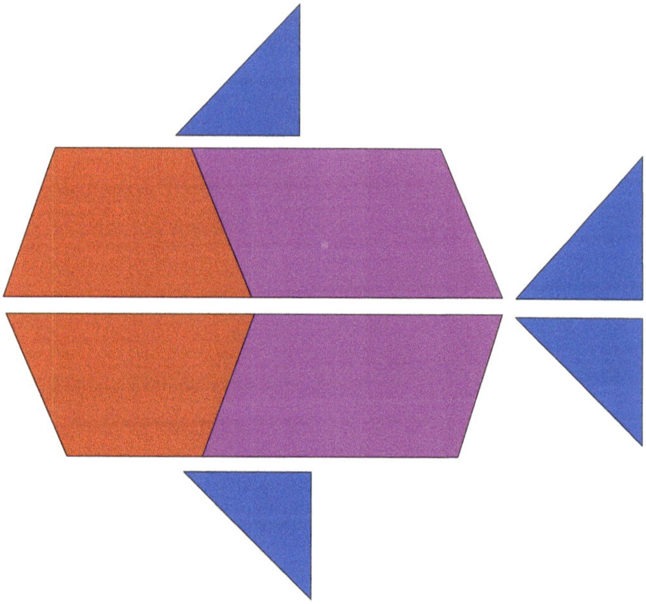

5. Repeat steps 3 and 4 to make lots of shapes in different colors.
6. Arrange some of your colored shapes on a piece of white paper. Follow the patterns shown here or design your own pictures.
7. If you want to keep a picture, glue the shapes onto the white paper.

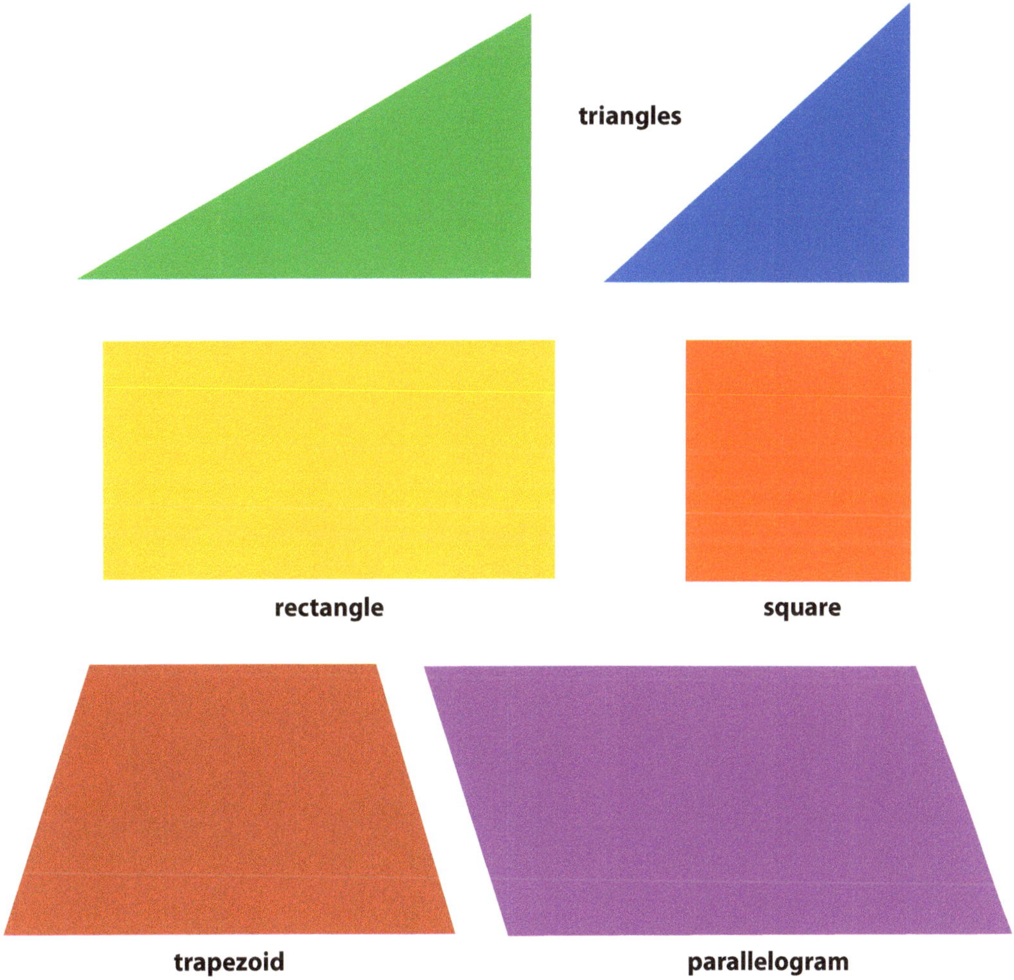

Decorate with Shapes

Some shapes fit neatly together. Look at the patterns on these pages. What shapes do you see?

A mosaic (moh ZAY ihk) is a decoration made of tiny pieces of clay, glass, stone, or other material. These pieces are called tesserae (TEHS uh ree). You may find mosaics in floors and walls. The ancient Romans liked mosaics. The floors and walls of Roman temples and villas were often covered with complicated patterns.

triangle

square

octagon

Try This...

The mosaic shown here is made of colorful stones. But you can make your own artwork out of colored paper or old magazines! Experiment to find out which shapes fit together. Glue your shapes onto a piece of cardboard to make a picture or a design.

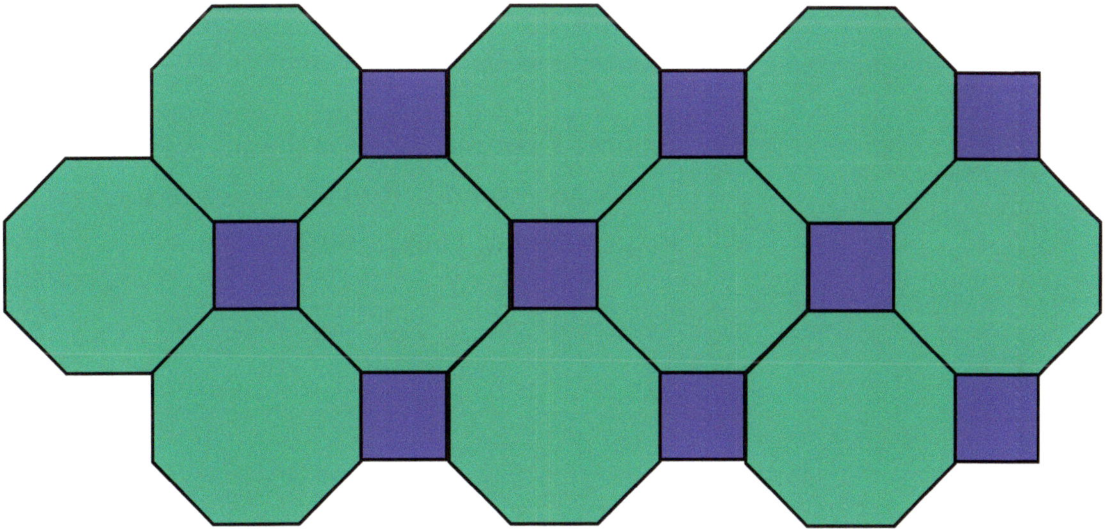

A party hat is a cone.

A soccer ball is a sphere.

A can is a cylinder.

Solid Shapes

As you know, some shapes are flat and some are not. Squares and circles are flat, but cubes, spheres, cylinders, and cones are not.

A flat shape has two dimensions: length and width.

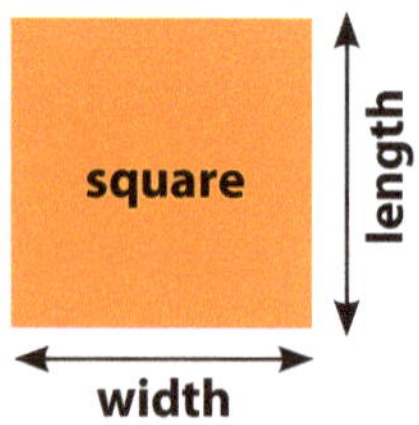

A solid shape has three dimensions: length, width, and depth (thickness).

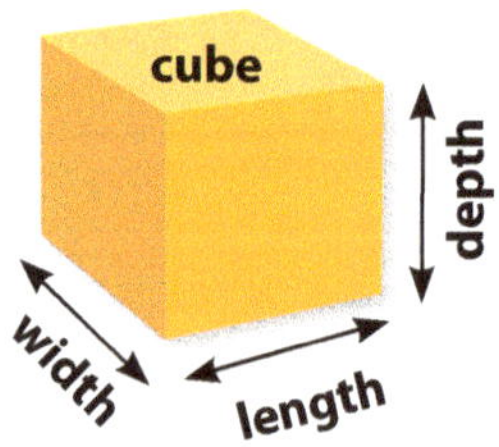

One example of a cube is a die.

Some examples of solid shapes are shown here. Can you find other examples of solid shapes around your home?

Try This...

Cube Puzzles

If you lay one cube on a table, you should be able to touch five of its sides. That means if you lay down two cubes, you'll be able to touch 10 sides, right?

Not always!

You Will Need:

- four cubes or dice

What To Do:

1. Find a way to lay down two cubes so that you can only touch eight sides.
2. Now try something a little harder: Lay down three cubes so that only 11 sides can be touched.
3. Here's one last puzzle: Can you lay down four cubes so that only 12 sides can be touched?

Check your solutions on page 179.

The Story of Numbers

What would we do without numbers?
Why, we couldn't even count!
We couldn't keep track of birthdays
or measure size or amount.

But have there always been numbers?
Were they always just lying around?
Were numbers a great discovery
that someone one day found?

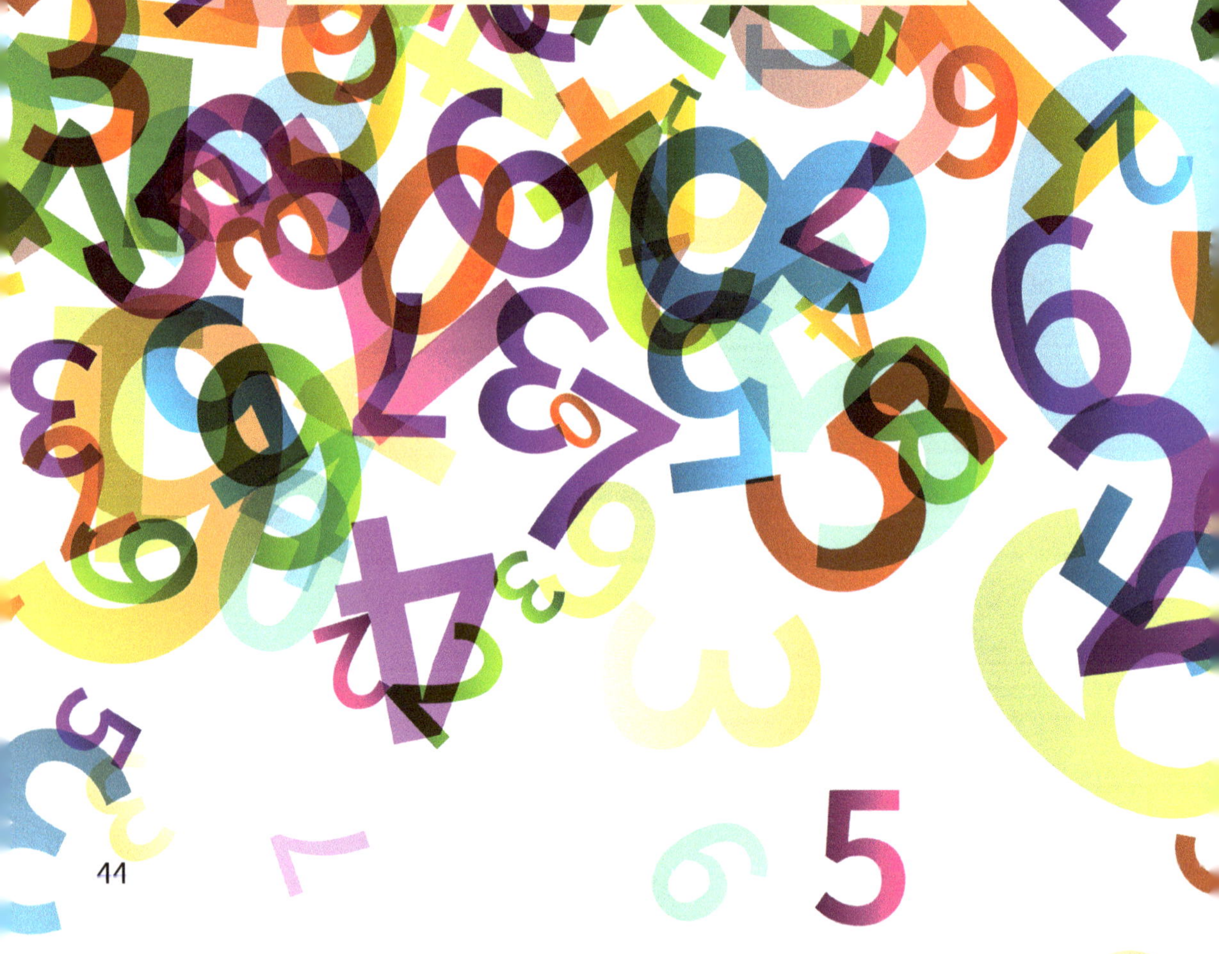

Numbers Everywhere

Think of all the numbers you use every day.

You read books (like this one) with numbered pages. Numbers describe the time and the channel of your favorite TV show.

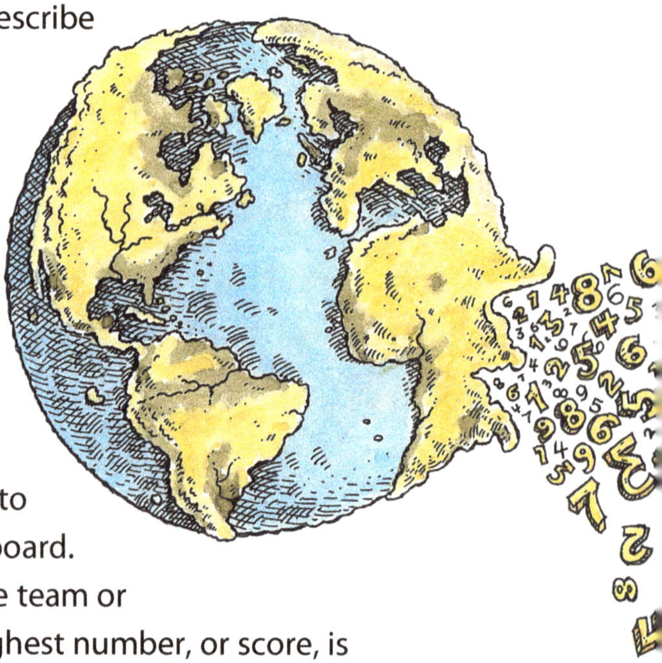

You probably play games that use numbers. You might roll dice or spin a spinner and count how many spaces to move on a game board. In many sports, the team or player with the highest number, or score, is the winner. In other sports, like golf, the lowest number is the winning score!

There are more numbers than you could count in an hour, a day, or even a lifetime. Think of everyone that lives in your city. They all have different telephone numbers! The street you live on may be long, but no other building on your street has exactly the same address, or number, as the one you live in.

Numbers also tell you a price, or how much something costs. They tell how much flour to use when you bake a cake. They tell how tall you are, how much you weigh, and how old you'll be on your next birthday! The world is truly full of numbers.

The First Ten

Long ago, people didn't need to use numbers. They didn't need to count! But over time, people's lives changed. They needed to count sheep, baskets of grain, and other important items.

The numbers 1 through 10 have special names in most languages because people learned to count by using their fingers. They would count up to 10 fingers and then start over.

In Chinese, the words for the numbers that come after 10 are based on the first 10 numbers.

Say the numbers 11 through 19.

Do you hear the "10" in each one? Shí yī means "ten and one." What does shí èr mean? Twenty means "two 10's" and thirty means "three 10's." What do you think forty means?

One hundred is a special name for ten 10's. One thousand is a special name for ten 100's.

Look on the next page at the chart of some number words in other languages. You can see that 1 through 10 are special numbers, and the rest are based on them.

	1	3	6	10	16	30
Spanish	uno	tres	seis	diez	dieciséis	treinta
French	un	trois	six	dix	seize	trente
German	eins	drei	sechs	zehn	sechzehn	dreissig
Arabic	wahid	tsalatsa	sitta	ashara	sittata ashar	tsalatsoun
Japanese	ichi	san	roku	juu	juuroku	sanjuu
Chinese	yī	sān	liù	shí	shí liù	sān shí

Pebbles for Tens

Miriam was only seven, but she worked hard like everyone else in her family. Her job was to watch the family's sheep.

Miriam's grandfather had watched the family's sheep when he was a boy. Back then, he counted the nine sheep on his fingers. Miriam used her fingers to count the sheep, too, but it was harder now. The flock had grown to nearly 50 sheep.

Every day Miriam counted the sheep. When she had touched all 10 fingers, she would say, "There's a 10 of sheep." Then she would count another 10 of sheep. But sometimes she forgot how many 10's she had counted. Then she had to start again. Miriam was a smart girl. She soon found a solution to her problem. Can you guess what she did?

Miriam gathered some pebbles. Each time she counted a 10 of sheep, she put a pebble on the ground. The last group of sheep came to eight fingers. There were four pebbles on the ground. That meant she had four 10's and eight fingers of sheep, or 48 sheep.

As Miriam learned, counting a lot of things is much easier when you use pebbles or other markers.

Pebbles in the Sand

Imagine that you are a rich merchant, living 3,500 years ago in Babylon in what is now the country of Iraq. You have 37 baskets of grain in your warehouse. You just bought 51 more baskets of grain from a farmer. You want to know how many baskets of grain you have now. How can you find out?

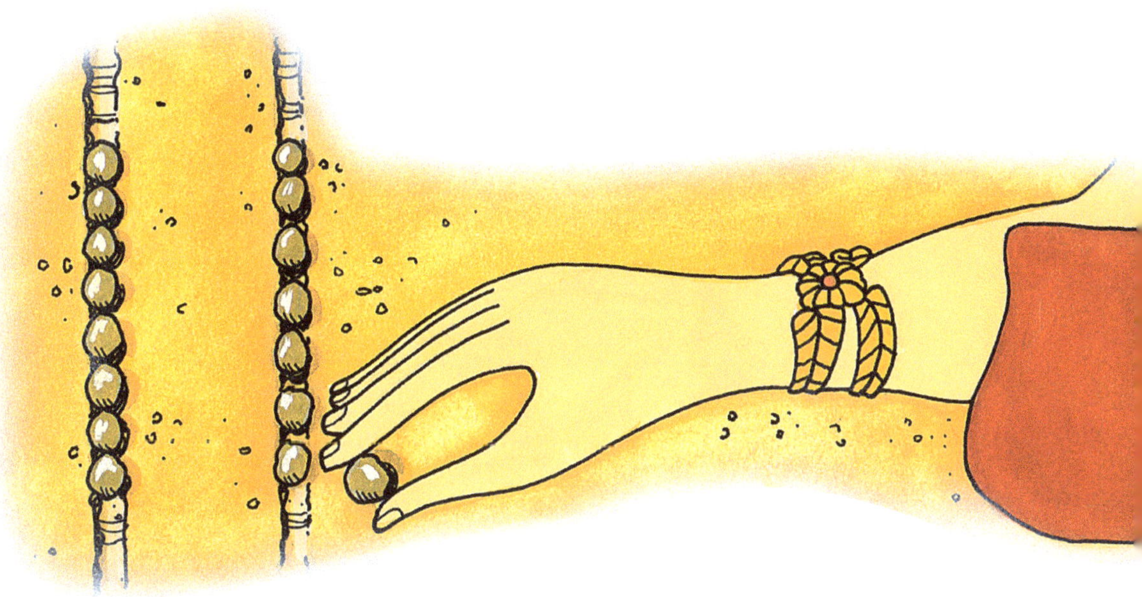

First you draw two grooves, or lines, in the sand with your finger. Then you place seven pebbles in the right groove. Each pebble stands for one basket of grain.

In the left groove you place three pebbles. Each pebble in the left groove stands for 10 baskets of grain.

The pebbles in the two grooves show that you have three 10's (30) plus seven 1's (7), or a total of 37 baskets of grain in your warehouse.

Now you add pebbles for the 51 baskets of grain you just bought.

You place one more pebble in the right (1's) groove and five more pebbles in the left (10's) groove.

Then you count the pebbles in each groove. There are eight pebbles in the 10's groove. Eight 10's **(10 + 10 + 10 + 10 + 10 + 10 + 10 + 10)** are 80. There are eight pebbles in the 1's groove. Eight 1's are 8. So you have **80 + 8**, or a total of 88 baskets of grain.

The Abacus

The Babylonian merchant's pebbles and grooves were an early form of a counting tool called an **abacus** (AB uh kuhs).

The abacus was invented by people in many parts of the world. The Babylonians and Egyptians used it 5,000 years ago. So did the Chinese.

The first abacus was probably a tray covered with dust or sand. People made counting marks with a finger and erased them with a sweep of the hand.

After a while, people began to make grooves in the sand and use pebbles for counters. Finally, they strung pebbles or beads on wires inside a wooden frame. That is what an abacus looks like today. It is easy to use and to carry around.

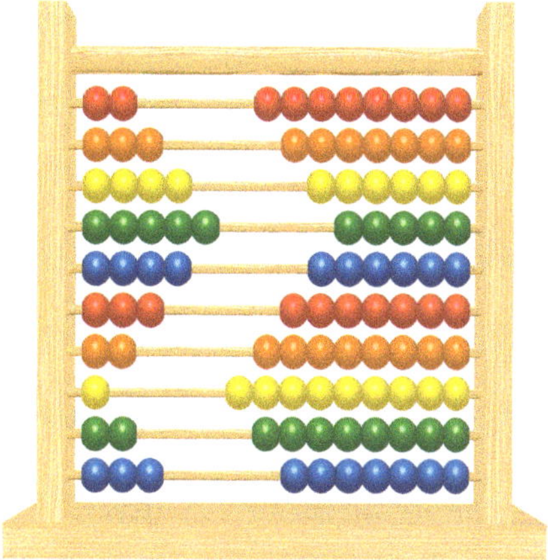

An abacus can be used to work all kinds of math problems. If you know how to use one, you can solve problems very quickly. In some contests, a person using an abacus has solved problems faster than a person using a calculator!

KNOW IT ALL

When you buy something in a store, you pay for it at the counter. Do you know how the counter got its name? Long ago, that's where merchants placed an abacus, or counting board. Today most stores have a cash register on the counter. But some shopkeepers in China and other parts of Asia still use an abacus.

Inventing Numerals

Once people began to count, how did they remember the numbers they had counted? They needed a way to write down the numbers—so they invented **numerals**. Numerals are symbols that stand for numbers.

Some of the first numerals were invented by the Egyptians about 5,000 years ago. Their marks for the first nine numbers look like pictures of fingers.

The Egyptians had special symbols for the numbers 10, 100, 1,000, 10,000, and 100,000. A picture of a lotus flower was the numeral for one thousand. There were thousands of lotus flowers in the Nile River near Egypt. So the Egyptians probably thought a lotus was a good symbol for a big number like 1,000.

1,000

The numeral for one hundred thousand was a picture of a tadpole. Lots and lots of frogs lived along the Nile River. When their eggs hatched, the water was probably full of tadpoles. That may be why the Egyptians used the tadpole as their symbol for a huge number like 100,000.

100,000

Count Like an Ancient Egyptian

This is what the Egyptian numerals looked like:

The finger stroke stood for 1.  **1**

The arch stood for 10.  **10**

The curved rope stood for 100. **100**

The lotus flower stood for 1,000.  **1,000**

Try This...

Get some paper and a pencil. Practice writing the Egyptian symbols shown on this page. Then see if you can write these numerals in Egyptian:

1. 35

2. 112

3. 1,245

4. 10,437

5. 162,354

Check your answers on page 180.

The bent finger stood for 10,000. 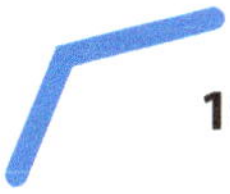 **10,000**

The tadpole stood for 100,000.  **100,000**

To show more than one of anything, the Egyptians repeated the symbol the correct number of times. The numeral below stood for 23:

But these numerals also meant 23:

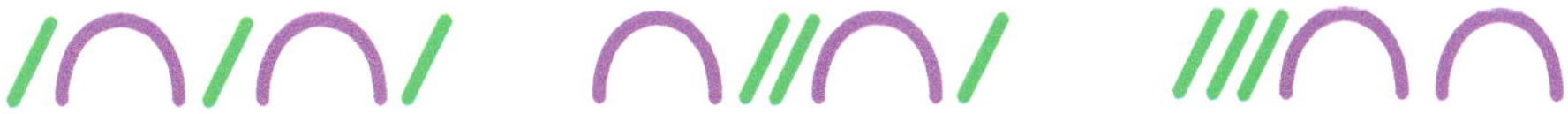

The Egyptians usually wrote their numerals from left to right. But sometimes they wrote them from right to left or from top to bottom.

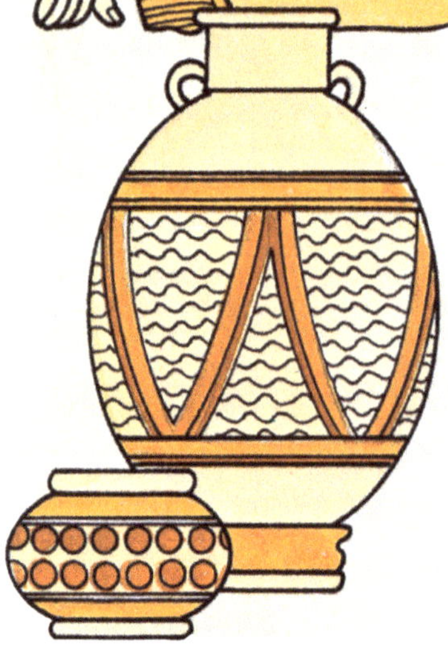

Other Kinds of Numerals

People all over the world invented different ways of writing numbers.

About 5,000 years ago, the Babylonians wrote on clay tablets with a little stick. The stick made a small mark shaped like an arrowhead. The Babylonian numerals for 1 to 10 looked like this:

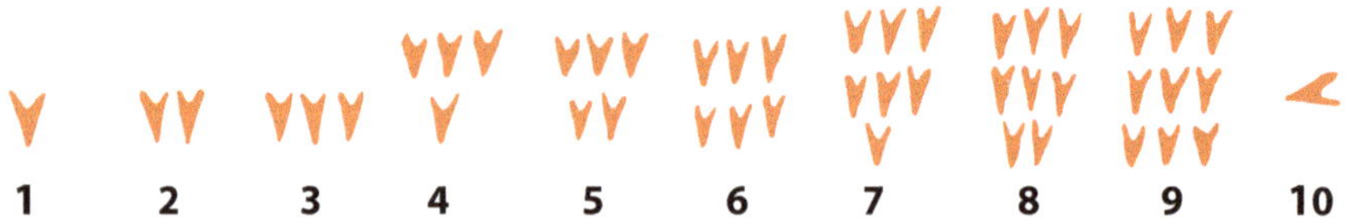

About 2,500 years ago, the ancient Greeks used the letters of their alphabet to stand for numbers. Their numerals for 1 to 10 looked like this:

A	B	Γ	Δ	E	F	Z	H	Θ	I
1	2	3	4	5	6	7	8	9	10

About 1,800 years ago, the Maya, an Indigenous people of what is now Central America and southern Mexico, developed a clever way of writing numbers. They must have used their fingers and their toes when counting because their number system is based on 20. The Maya numerals for 1 to 10 looked like this:

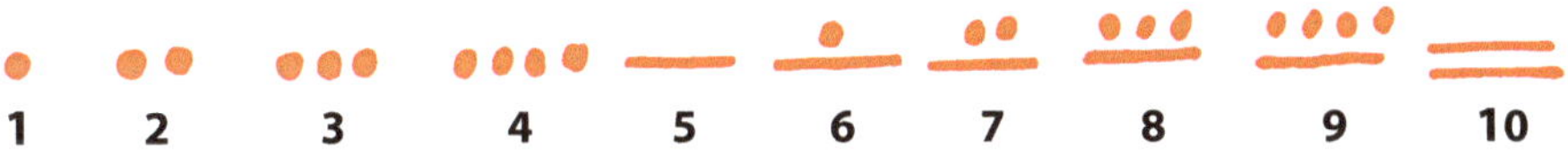

About 2,000 years ago, the Hindu people of India were using numerals that looked like this:

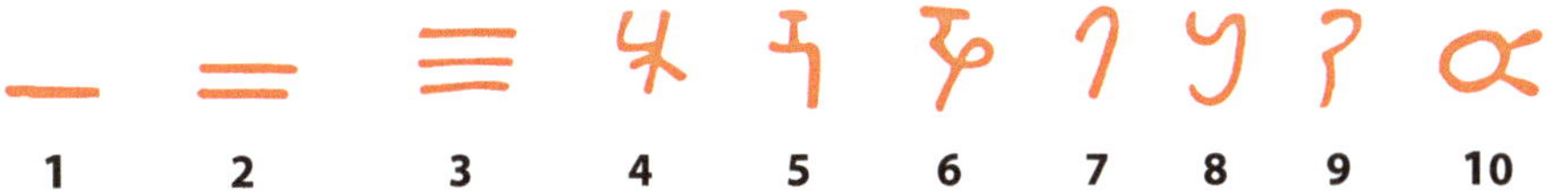

These Hindu numerals are very important to us. Over many years, and with some changes, they became the numerals we use today.

Roman Numerals

Can you read the numbers on the clock at the bottom of this page? They are called Roman numerals. They were invented by the ancient Romans around 500 B.C.

The Romans made the numerals for one, two, three, and four look like fingers. The Roman numeral for five looks like the space between your thumb and first finger when your hand is spread open. The Roman numeral for 10 looks like two crossed hands.

After a while, the ancient Romans found a way to save time and space when writing numerals—**subtraction**! They used subtraction to make new symbols for the numbers four and nine. IIII became IV, and VIIII became IX.

I II III IV V VI

The new symbols followed a rule: When the first numeral is smaller than the second numeral, subtract the smaller one from the larger one.

IV = V– I (5 – 1) **IX = X – I (10 – 1)**
IV = 4 **IX = 9**

The symbols I, V, and X are not just numerals. They are also letters in the Roman alphabet. Other letters stand for larger numbers: L = 50, C = 100, D = 500, and M = 1,000.

VII VIII IX X XI XII

Try This...

When in Rome...

Let's count like the ancient Romans!

Remember these rules:

When the Romans wrote a smaller number to the left of a larger number, they subtracted the smaller amount. So **IX** means **10 – 1**, or **9**.

When the Romans wrote a smaller number to the right of a larger number, they added the two together. So **XI** means **10 + 1**, or **11**.

You Will Need:

- paper
- a pencil

What To Do:

Write the missing numerals.

1. Counting from 1 to 10:
I, ______, III, IV, V, VI, ______, VIII, IX, ______

2. Counting by 10's to 150:
______, XX, XXX, XL, ______, LX, LXX, ______, XC, C, ______, CXX, CXXX, ______, CL

3. Now write these numbers in Roman numerals:

   a. 17

   b. 163

   c. 1,528

Roman numerals are still used today. You can see them on many clocks and watches. Sometimes people use them to record the year in which a movie was filmed or a book was published. Where can you find Roman numerals?

Check your answers on page 180.

Hindu-Arabic Numerals

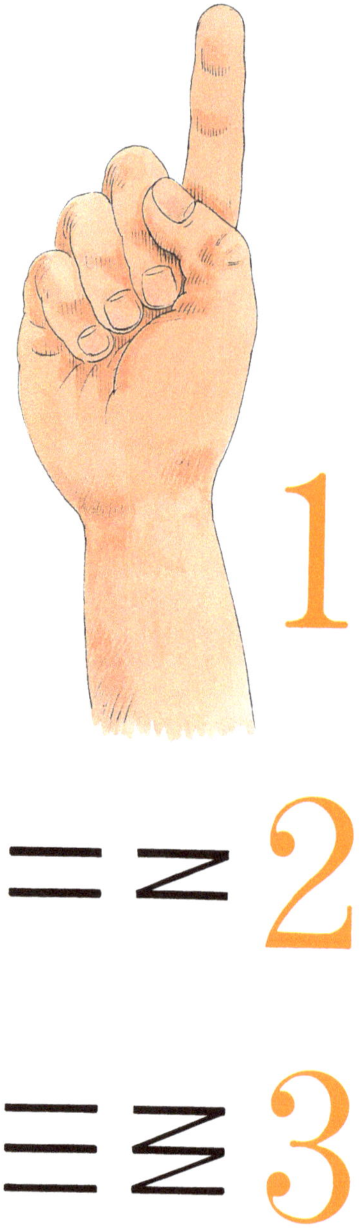

The Hindu numeral 1 probably stands for one finger. The numeral 2 may have started as two straight lines that were later connected. Can you see three straight lines in the numeral 3?

The Hindu numerals were probably invented about 2,000 years ago.

When the Arabs conquered Spain about 1,300 years ago, they brought the Hindu numerals with them. Because the Arabs brought these numerals to other lands, they became known as Hindu-Arabic numerals.

When the Arabs conquered Spain, the people in Europe were still using Roman numerals. For several hundred years, they went right on using them. Most Europeans did their arithmetic with an abacus. When you use an abacus, it doesn't matter what written number system you use.

But over time, European mathematicians found that it was easier to write math problems with Hindu-Arabic numerals than with Roman numerals. That's because Hindu-Arabic numerals have **place value**.

To understand place value, look at these Hindu-Arabic numbers:

In 25, the 5 stands for five 1's, or 5.

In 51, the 5 stands for five 10's, or 50.

In 539, the 5 stands for five 100's, or 500.

A Hindu-Arabic numeral can mean 1's, 10's, 100's, or even more, depending on the place where it is written. But Roman numerals, such as V, have the same meaning no matter where they are written in a string of numbers.

After European mathematicians started using Hindu-Arabic numerals, other people in Europe began to use them, too. This is how the Hindu-Arabic numerals looked about a thousand years ago, when they were becoming popular in Europe:

Zero

One of the most important things ever invented was—nothing!

How can "nothing" be important? And how can "nothing" be invented? To understand, let's think about the abacus.

To show the number 105 on an abacus, you push up one bead in the 100's column, no beads in the 10's column, and five beads in the 1's column.

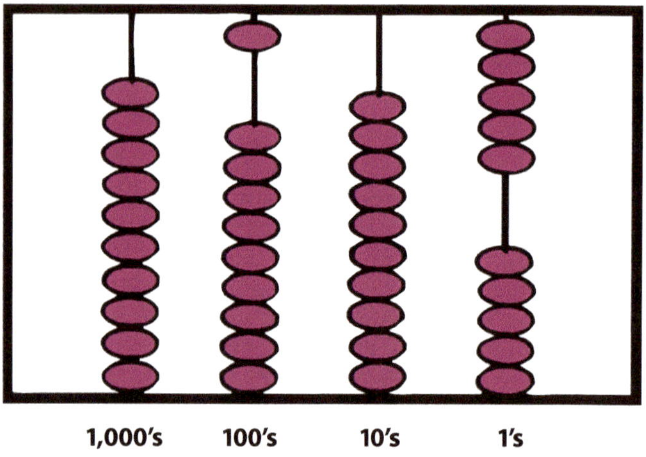

The Egyptian and Roman numeral systems had no zero. To write a number such as 105, people used the symbols for 100 and for 5. But the Hindu number system, like the abacus, had place value. The Hindus needed a way to show "nothing."

Arabic Egyptian Roman

By about A.D. 800, the Hindus had invented a numeral called sunya. It stood for an empty wire on an abacus. With just nine numerals and sunya, the Hindus could write any number!

The Arabs used the new numeral along with the other Hindu numerals. They gave it a new, Arabic name: sifr. Europeans adopted the new numeral along with the rest of the Arabic numerals. In fact, the English word *zero* comes from the Arabic numeral sifr, meaning "empty."

The Arabs spread the Hindu numerals, including zero, across Europe.

By about A.D. 250, the Maya people in Central America also had invented a way to show "nothing." They used several different symbols for zero. One of the symbols looked like a snail's shell:

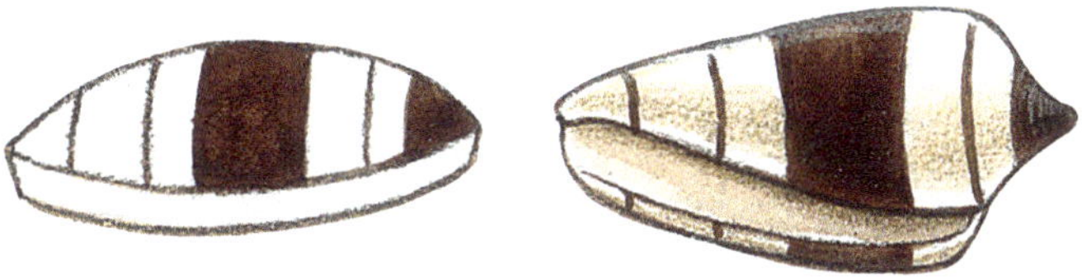

Other Ways to Count

In some parts of the world, people use number systems that are not based on 10.

The Mayan number system was based on 20. We can see this by looking at some Mayan number words:

1	hun
2	ca
5	ho
10	lahun
20	hun kal
40	ca kal
100	ho kal

Notice that the Mayan word for 20, hun kal, means "one 20." The word for 40, ca kal, means "two 20's." The word for 100, ho kal, means "five 20's." (Remember that in English, 40 means "four 10's" and 100 means "ten 10's.")

Some groups of people in Papua New Guinea, a nation in the Pacific Ocean near Australia, don't count on just their fingers and toes—they use their whole bodies! They point to the fingers, wrists, elbows, shoulders, ears, eyes, nose, and other parts of the body as they count.

Some groups in Papua New Guinea count 68 body parts. The Oksapmin people of Papua New Guinea use 27 body parts when counting, as shown below. Others count 33 or some other number. The body parts are usually counted in the same order.

The ancient Babylonians used a number system that was based on 60. The reason that there are 60 seconds in a minute and 60 minutes in an hour is because of their system!

Making Numbers Count

You can have lots of fun with numbers!
You can put them in groups or in pairs.
You can use them to solve a puzzle,
or arrange them in magic squares.

You can add or subtract them forever.
You can make them very small.
In fact, you can do, with a number or two,
almost anything at all!

Sorting with Circles

A **Venn diagram** is a picture that shows how things are alike and how they are different.

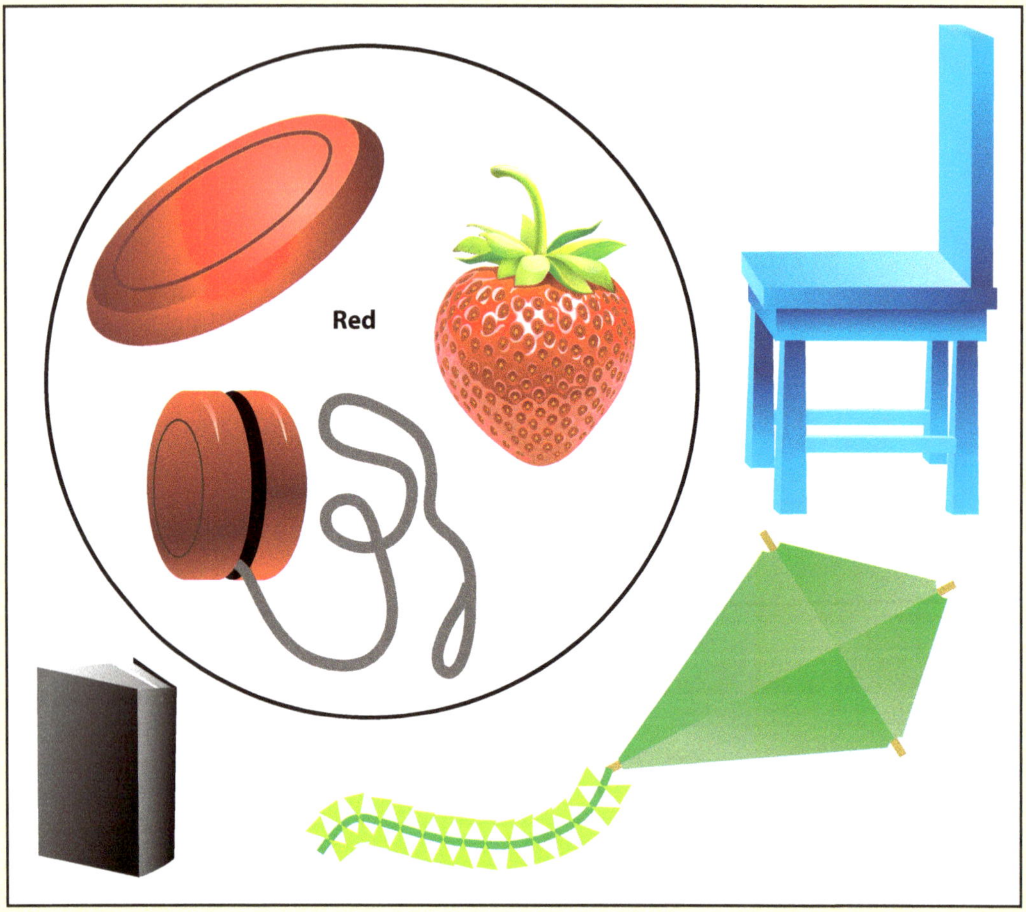

This Venn diagram shows things that are red and things that are not red. Everything in the circle is red. Everything outside the circle is not red.

The Venn diagram above shows things that are pink and things that are brown. Everything in the left circle is pink. Everything in the right circle is brown. Everything outside the circles is neither pink nor brown.

This Venn diagram shows things that are green and things that are yellow. Everything in the left circle is green. Everything in the right circle is yellow. Everything outside the circles is neither green nor yellow.

Look carefully at the middle of this diagram! The green and yellow circles overlap. In the overlapping part, called the intersection (IHN tuhr SEHK shuhn), we see things that are green and yellow.

Try This...

Toy Pile-Up

Take a look in your toy box or closet. It's probably full of fun things to play with, like stuffed animals, dolls, toy cars, and balls. Not all your toys look alike. That's easy to see! But maybe they are alike in some ways. Play this game and find out!

You Will Need:

- 5-10 different toys
- a friend to play with

What To Do:

1. Put the toys in a pile on the floor.

2. Think of a word or phrase that describes how some of the toys are alike, such as big, round, yellow, or soft. Tell your friend how to sort the toys. ("Find all the balls.")

3. Your friend sorts the toys into piles and tells you the results. ("There are two balls. There are five toys that aren't balls.")

4. Now, your friend tells you how to sort the toys. ("Find all the animals.") Then, sort the toys.

5. Continue until you can't think of any more ways to sort the toys! How many ways did you find?

6. Put your toys away when you're done!

Try This...

Scavenger Hunt

Now that you have read about Venn diagrams and have practiced sorting things, it's time to combine the two.

You Will Need:

- 3 pieces of paper
- colored pencils or crayons
- a clock or timer (optional)

What To Do:

1. Draw a circle on one piece of paper. Inside the circle, write Toys.
2. Draw two circles on another piece of paper. Write Foods inside one circle and Drinks inside the other circle.

3. Draw two overlapping circles on the third piece of paper. Write Orange inside one circle and White inside the other circle.

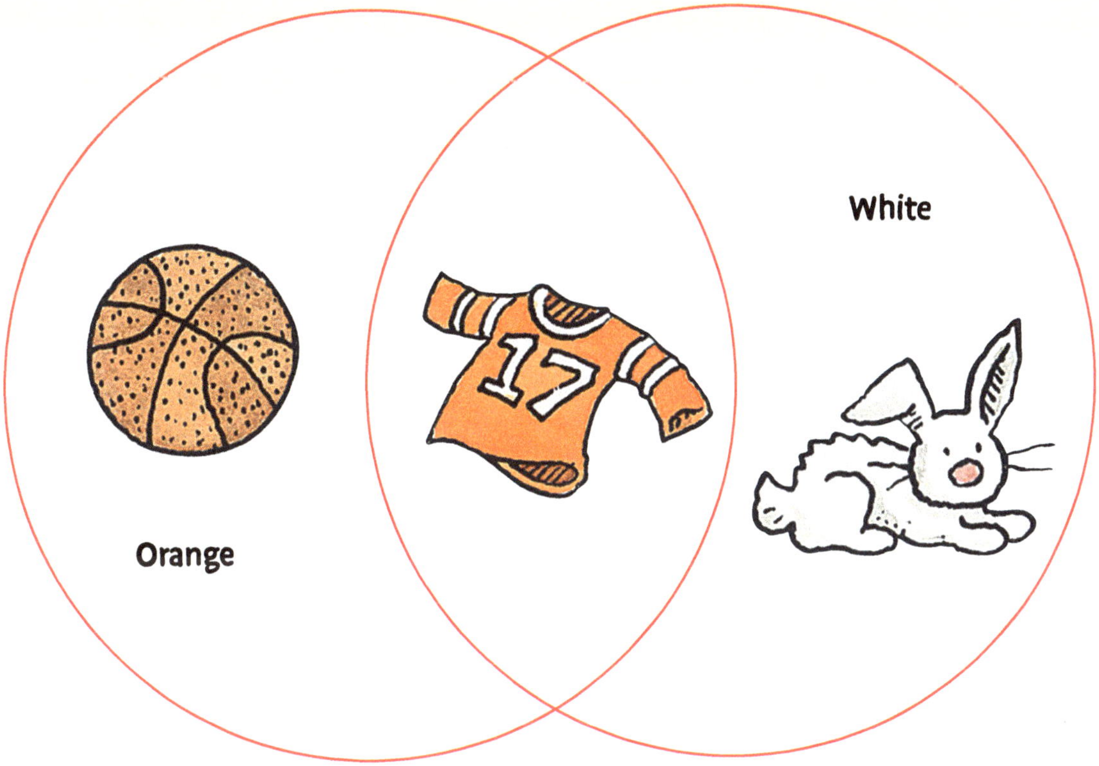

4. Look around your home for objects that belong on each page. You can draw pictures of the objects or write their names in the circles. Don't use an object on more than one page!

5. For a bigger challenge, ask an adult to time you. How many objects can you find for your Venn diagrams in five minutes?

Pretty Patterns

Mr. Green is watering his garden. Look at the flowers! There are red tulips, yellow daffodils, and purple irises.

Mr. Green planted his flowers in a pattern. Start at the left side of the row of flowers above. Follow the colors with your finger: red, yellow, purple, red, yellow, purple, red, yellow, purple.

The pattern of the flowers is red, yellow, purple. Do you see how the pattern repeats?

Mr. Green used a different pattern beside his fence. Follow the colors from left to right. The pattern is purple, yellow, red, yellow.

Now look at the flowers next to Mr. Green's front porch. What pattern do they follow?

You can check your answer on page 180.

Try This...

Draw your own flower garden picture with crayons or markers. Follow a pattern as you "plant" the flowers. Then show your picture to a friend. Ask your friend to describe the pattern of the flowers.

Try This...

Jewelry to Wear—and Eat

Now you can make a pattern that tastes as good as it looks!

You Will Need:

- licorice whips
- O-shaped cereal
- O-shaped hard candies
- tube-shaped licorice candies
- a piece of paper

What To Do:

1. Look at the necklace on these pages. It is made of cereal and candy. The pieces follow this pattern:

- 1 licorice candy
- 2 hard candies
- 3 cereal O's

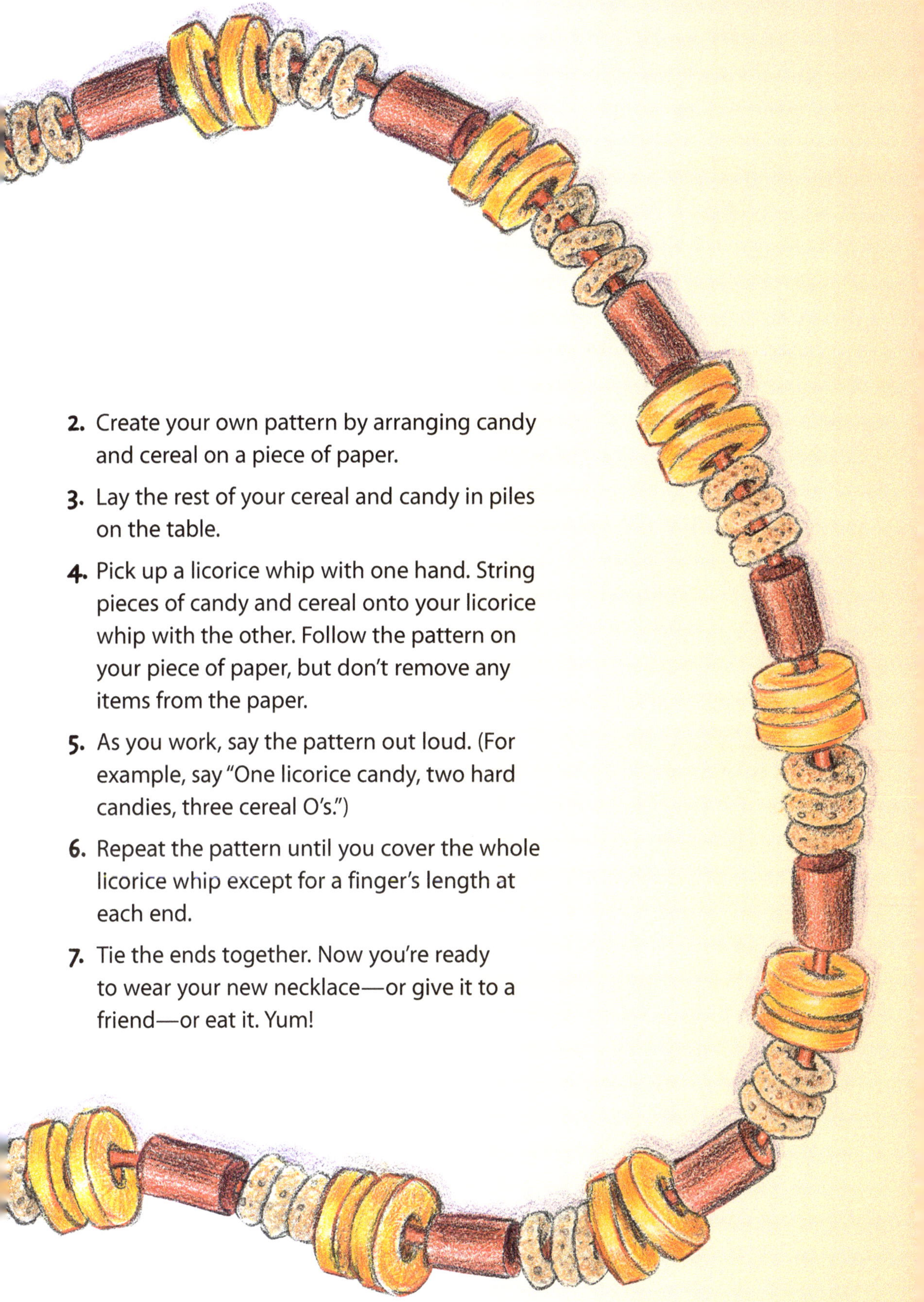

2. Create your own pattern by arranging candy and cereal on a piece of paper.
3. Lay the rest of your cereal and candy in piles on the table.
4. Pick up a licorice whip with one hand. String pieces of candy and cereal onto your licorice whip with the other. Follow the pattern on your piece of paper, but don't remove any items from the paper.
5. As you work, say the pattern out loud. (For example, say "One licorice candy, two hard candies, three cereal O's.")
6. Repeat the pattern until you cover the whole licorice whip except for a finger's length at each end.
7. Tie the ends together. Now you're ready to wear your new necklace—or give it to a friend—or eat it. Yum!

The Art of Addition

Kelly and Nick are playing with toy cars. Kelly has five cars. Nick has three cars. How many cars do they have in all?

To find out, we use **addition** (uh DIHSH uhn). Addition means putting numbers or groups of things together.

One way to add is by "counting on." You know that Kelly has five cars. So you think, "five." Then you point to each of Nick's cars and "count on" from five: "six, seven, eight." Together, Kelly and Nick have eight cars.

You can write an addition statement like this:

$$\mathbf{5 + 3 = 8} \text{ or } \begin{array}{r} \mathbf{5} \\ \underline{\mathbf{+\,3}} \\ \mathbf{8} \end{array}$$

We say, "Five plus three equals eight."

Find eight small objects, such as toy cars, blocks, or checkers. Put five of them in a group. Put three in another group. Now add the groups together. Write the addition statement: **5 + 3 = 8**.

Put your objects in a group of seven and a group of one. Add the groups together. Write the addition statement: **7 + 1 = 8**

Find other ways to group your objects. Write the addition statements. How many simple addition statements can you make by grouping eight objects?

You can check your answers on page 180.

Patterns in Addition

Basic statements in addition are called addition facts. When you learn the addition facts, you will be able to add small groups of things quickly, without "counting on."

Here are some addition facts:

2 + 1 = 3

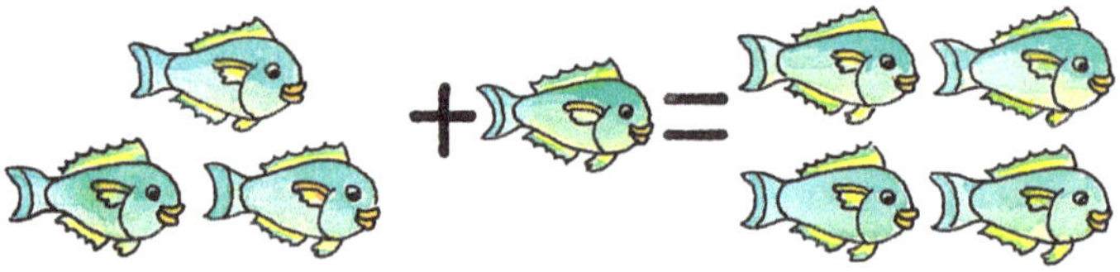

3 + 1 = 4

4 + 1 = 5

Do you see a pattern? Try following the pattern to complete these addition facts.

5 + 1 = ___

6 + 1 = ___

7 + 1 = ___ 

8 + 1 = ___ 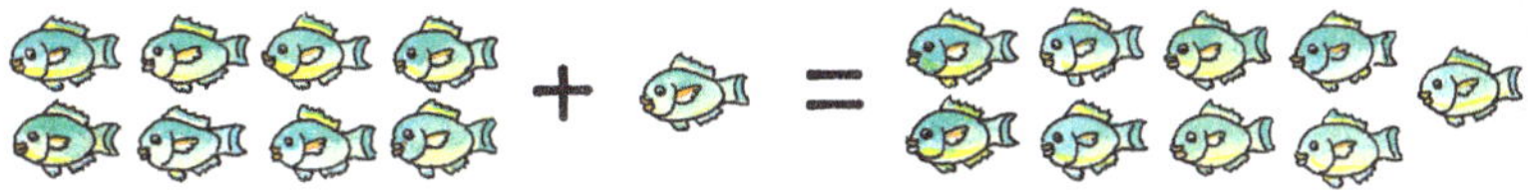

9 + 1 = ___ 

When you add 1 to any number, you get the next higher number.

Now look at these addition facts.

1 + 2 = 3

1 + 3 = 4

1 + 4 = 5

They are very similar to the first set of facts, aren't they? The first two numbers in each fact are switched. That's the only difference. The answer stays the same!

When you add numbers or groups of objects, you can put them in any order. You will always get the same answer.

You can check your answers on page 180.

Addition Facts

There are 100 basic addition facts. That's a lot, isn't it? But you can learn them all easily if you look for patterns in addition.

Here is a table of basic addition facts. To find the answer to a problem like **1 + 3**, look across the row that starts with 1 and down the column that starts with 3. Where the row and column meet, you will find the answer to the addition problem.

Have fun looking for patterns in the table!
Here are some questions to get you started:

1. What happens when you add 0 to a number?

2. Where can you find the answers to problems like **1 + 1**, **2 + 2**, **3 + 3**, and so on?

You can check your answers on page 181.

+	0	1	2	3	4	5	6	7	8	9
0	0	1	2	3	4	5	6	7	8	9
1	1	2	3	4	5	6	7	8	9	10
2	2	3	4	5	6	7	8	9	10	11
3	3	4	5	6	7	8	9	10	11	12
4	4	5	6	7	8	9	10	11	12	13
5	5	6	7	8	9	10	11	12	13	14
6	6	7	8	9	10	11	12	13	14	15
7	7	8	9	10	11	12	13	14	15	16
8	8	9	10	11	12	13	14	15	16	17
9	9	10	11	12	13	14	15	16	17	18

Try This...

You Can Add!

Read each problem below. Put counters in the small bowls to show the numbers you need to add. Then move all the counters into the third bowl. Add the counters in the third bowl to find the answer to the problem.

You Will Need:

- 20 buttons, coins, or other counters
- 2 small bowls or boxes
- a bigger bowl or box

What To Do:

1. Kyoko has 4 video games. Her aunt gives her 1 video game for her birthday. How many video games does Kyoko have now?
2. Paul is at the library. He wants to check out 6 books about whales and 3 books about the planets. How many books does he want to check out?

3. There are 7 children playing soccer in the park. Then 5 more children join the game. How many children are playing soccer in all?

4. There are 11 girls and 9 boys in Yolanda's class. How many children are in the class?

You can check your answers on page 181.

Magic Squares

You can do magic with addition! Look at the square in the picture below. It is called a magic square.

When you add up any line of numbers in a magic square, you get the same answer. Try it yourself and see!

The numbers in the top row:
4 + 9 + 2 = ___

The numbers in the middle row:
3 + 5 + 7 = ___

The numbers in the bottom row:
8 + 1 + 6 = ___

The numbers in the three columns, top to bottom:
4 + 3 + 8 = ___
9 + 5 + 1 = ___
2 + 7 + 6 = ___

The numbers in the diagonal lines:
4 + 5 + 6 = ___
2 + 5 + 8 = ___

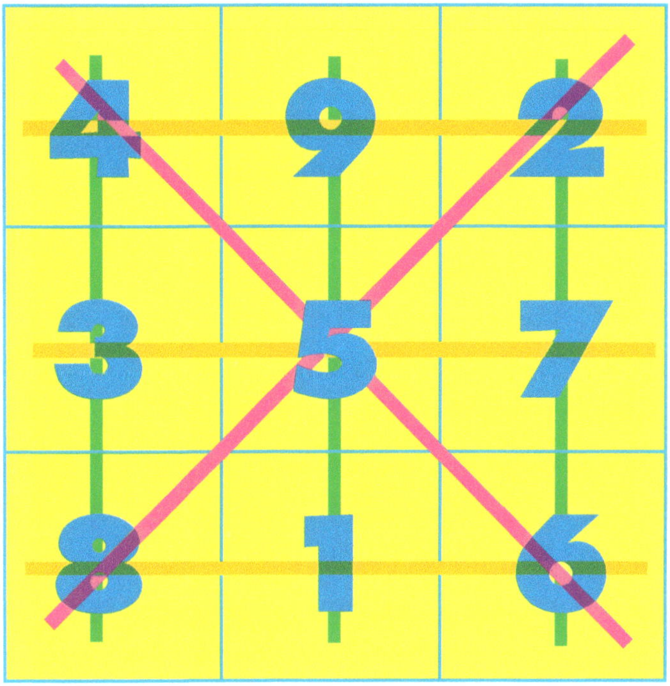

What's the magic number for this square? That's right—it's 15!

Now write the numbers from this square in order from smallest to biggest.

You've just discovered another important fact about magic squares! The numbers in a magic square are always consecutive (kuhn SEHK yuh tihv). That means they follow one after the other in order.

KNOW IT ALL

If you know how to multiply (see pages 110-111), you can find the "magic number" for any 3 x 3 magic square. Just multiply the center number by 3. For the magic square on this page, 5 x 3 = 15.

More Magic Squares

Here is a famous magic square by the German artist Albrecht Dürer. It is a 4 x 4 square. When you add up the lines of numbers going across, down, and diagonally, you get the same answer—34. Try it and see for yourself.

That's not all the magic in Dürer's magic square. Add up the numbers in the four corners. Then add up the four numbers in the center. What do you get? Right! You get 34 each time!

There are at least six other ways to get 34 by adding four numbers from the square. See if you can find them!

16	3	2	13
5	10	11	8
9	6	7	12
4	15	14	1

You can find the answers on page 181.

Try This...

Can you complete these magic squares?
Find the missing numbers!

3	-	13	0
-	5	-	11
4	-	10	-
15	2	-	12

5	-	-
6	4	2
-	8	-

You can check your answers on page 182.

Try This...

The Magic Number in Your Name

Long ago, people believed numbers were magical. They made up ways of using numbers to tell them things. You can use addition to find your own "magic" or lucky number! Here's how.

You Will Need:

- paper
- a pencil

What To Do:

1. Print your whole name on the piece of paper. Below each letter, write the number shown on this chart:

A = 1	B = 2	C = 3	D = 4	E = 5	F = 6
G = 7	H = 8	I = 9	J = 1	K = 2	L = 3
M = 4	N = 5	O = 6	P = 7	Q = 8	R = 9
S = 1	T = 2	U = 3	V = 4	W = 5	X = 6
Y = 7	Z = 8				

A	R	C	H	I	B	A	L	D
1	9	3	8	9	2	1	3	4

A	R	T	H	U	R
1	9	2	8	3	9

A	L	A	D	D	I	N
1	3	1	4	4	9	5

2. Add up the numbers. (To make it easier, you may want to add them separately for each word, then add the word totals together.) You will get an answer with two or three numerals, such as 99 or 111.
3. Add the numerals together. If you get a number from 1 to 9, that's your magic number. If you get a bigger number, add those two numerals together to find your magic number. For example, if you get 13, add 1 and 3 to get 4 as your magic number.

Archibald Arthur Aladdin's numbers add up to 99. When he adds 9 and 9, he gets 18. Then he adds 1 and 8 to get 9. His magic number is 9. What's your magic number?

The Science of Subtraction

Tanisha saw six birds in a tree. Two of them flew away. Then how many birds were left in the tree?

Four birds are left in the tree. Do you see them?

Subtraction (suhb TRAK shuhn) means finding the difference between two numbers or two groups of things. Subtraction is the opposite of addition.

Look at the birds again. The difference between six birds (in the tree) and two birds (that flew away) is four birds (left in the tree).

You can write this as a subtraction problem:

$$\mathbf{6 - 2 = 4} \text{ or } \begin{array}{r} \mathbf{6} \\ \underline{\mathbf{-2}} \\ \mathbf{4} \end{array}$$

We say, "Six minus two equals four."

Basic statements in subtraction are called subtraction facts. Here are some subtraction facts:

3 – 1 = 2   **4 – 1 = 3**   **5 – 1 = 4**

Do you see a pattern? Compare the subtraction facts with these addition facts:

2 + 1 = 3   **3 + 1 = 4**   **4 + 1 = 5**

Each basic subtraction fact is related to an addition fact:

2 + 1 = 3   **3 – 1 = 2**

Try This...

You Can Subtract!

Read each problem. Put counters in the big bowl to show how many items you start with. Then move (subtract) the correct number of counters into the small bowl. Count the number of items left in the big bowl to find the answer to the problem.

You Will Need:

- 20 buttons, coins, or other counters
- a small bowl or box
- a big bowl or box

What To Do:

1. Kyoko has 5 video games. She gives 2 of them to her younger brother. How many video games does Kyoko have now?
2. Raul has 9 library books at home. He has read 3 of them. How many books are left to read?

3. There were 12 children playing basketball in the park. Now 4 have to go home for supper. How many children are still playing basketball?

4. There are 20 children in Yolanda's class. Today, 1 is absent. How many children are in the classroom?

You can check your answers on page 182.

Backward Puzzles

Most number puzzles ask questions. Here is a different kind of number puzzle. We give you the answers, and you figure out how to get them!

The rules are simple:

- Use each number only once.
- You may add and subtract to get the answer.

Are you ready? Try this one:

Puzzle 1

The answer is 6.

Use any three numbers from 1 to 5.

Here is one solution:

3 + 5 – 2 = 6

(First add 3 + 5 to get 8.
Then subtract 2 to get 6.)

What other solutions can you find?

Now try these puzzles.
How many ways can you solve each one?

Puzzle 2
The answer is 8.
Use any three numbers from 1 to 10.

Puzzle 3
The answer is 3.
Use any three numbers from 1 to 7.

You can see possible solutions on page 182.

The Minus Mystery

Mildred Minuette wrote mystery stories. Her books were very popular. Her fans begged for autographs at all hours of the day and night. Mildred was so busy signing autographs that she didn't have time to write more stories!

One day, Mildred disappeared. No one knew where she was. Then Penelope Puzzle, Mildred's best friend, got a note in the mail. It said:

Gone to **10 – 4, 20 – 2, 5 – 4, 20 – 6, 9 – 6, 8 – 3.**
Working on **9 – 7, 19 – 4, 20 – 5, 14 – 3.**
Don't **26 – 3, 16 – 1, 20 – 2, 19 – 1, 26 – 1.**
Back **20 – 1, 17 – 2, 18 – 3, 19 – 5.**

Where was Mildred, and what was she doing? Help Penelope solve the mystery.

First, subtract the numbers in the message and write the answers on a piece of paper.

Now you have a message in code. What do the numbers stand for?

If you need help, read the hint below.

You can check your answers on page 182.

What's the Biggest Number?

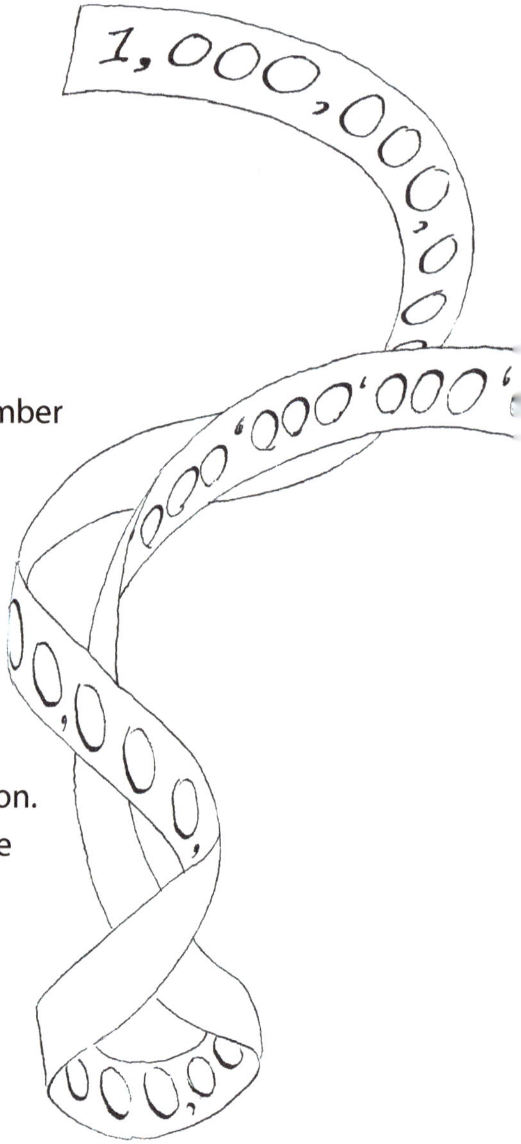

Do you know what the biggest number is? Is it 1 million? That's a one followed by six zeros—1,000,000.

No, 1 million isn't the biggest number. There can be many millions. For example, Earth is about 93 million miles (150 million kilometers) from the sun.

One million millions make a trillion. That's a big number! We write it like this: 1,000,000,000,000.

But 1 trillion isn't the biggest number, either. There's a much bigger number called a googol (GOO gahl). A googol is a one with a hundred zeros after it.

Is a googol the biggest number? No. There is no biggest number! You could write a number followed by millions and millions of zeros, but someone could make a bigger number just by adding a zero to your number!

When you say the alphabet, you know when to stop. But when you count, you could go on forever if you had names for all the numbers! That's because numbers are **infinite** (IHN fuh niht). They never come to an end.

Try This...

Ask your family and friends to help you collect large numbers of something. Choose an item that's small, inexpensive, and easy to find, such as buttons, marbles, or pebbles.

First try to collect 100 items. How much space do they fill? Then try to collect a thousand of them. It may take a long time and a lot of help to collect that many things! But your collection will help you see how big numbers really can be.

Go Figure!

The more we learn,
the more we know,
and numbers help
our knowledge grow!

We multiply
and then divide;
we're calculating
far and wide.

Numbers are
a magic way
to fill the hours
of any day!

The Magic of Multiplication

Imagine that you are playing a game with dice. You roll five 4's:

You could add the 4's to find out how many points you have:

4 + 4 + 4 + 4 + 4 = 20

But once you learn that five 4's make 20, you know how to multiply! You can write the problem like this:

5 x 4 = 20

We say, "Five times four equals twenty," or "Five fours are twenty."

Basic statements in **multiplication** (muhl tuh pluh KAY shuhn) are called multiplication facts. They're easy to learn if you know how to add.

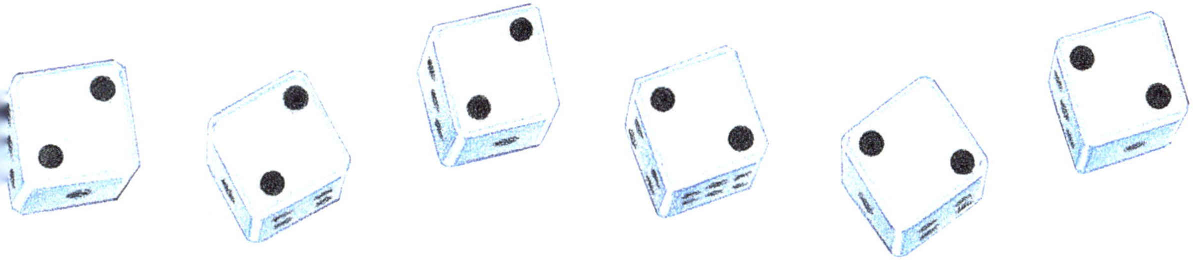

Here are two more multiplication facts:

6 x 2 = 12
2 x 6 = 12

Six 2's are 12, and two 6's are also 12.

Try the addition yourself:

2 + 2 + 2 + 2 + 2 + 2 = ?
6 + 6 = ?

When you multiply two numbers, it doesn't matter which one you write first. Look at the eggs in this carton. We can think of them as six rows of 2 or as two rows of 6. No matter which way you arrange the numbers, there are 12 eggs in the carton!

Counting Squares

Ahmose, chief builder for the king of Egypt, had a problem. He was building a new palace for the king and queen. The queen's bedroom floor was to be made of special white stone blocks from the land of Punt, an area on the northeast coast of Africa.

Ships would have to go for the stone blocks and bring them back. So Ahmose needed to know exactly how many blocks to order. If he didn't order enough blocks, the ships would have to go back for more—and the king would be angry.

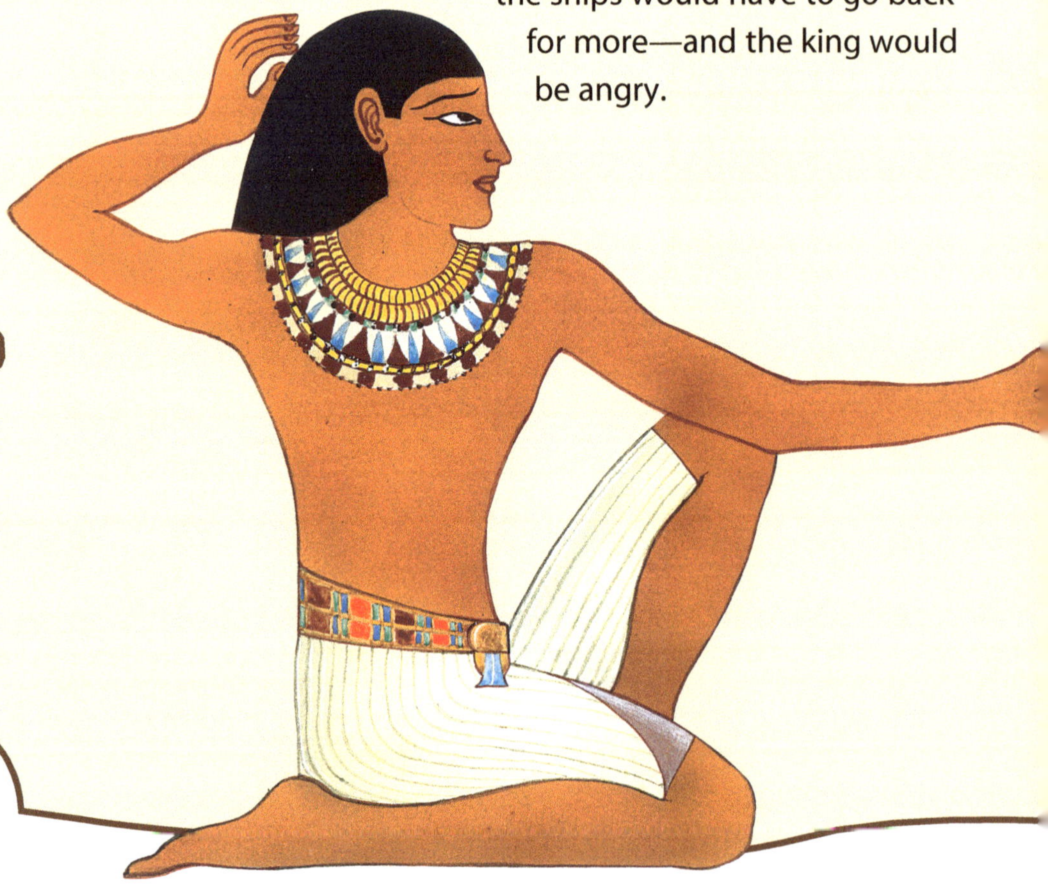

But ordering too many blocks would be a waste of money—and the king would be angry about that, too.

The queen's bedroom was to be 10 paces long and 9 paces wide. Ahmose decided to use square blocks that measured 1 pace on each side. That way, 10 blocks would be the exact length of the room. Nine blocks would be the exact width of the room.

On a piece of papyrus, which is an ancient kind of paper, Ahmose drew 9 rows, with 10 squares in each row. Then he carefully counted all the squares. There were 90 squares—so Ahmose knew he needed 90 stone blocks!

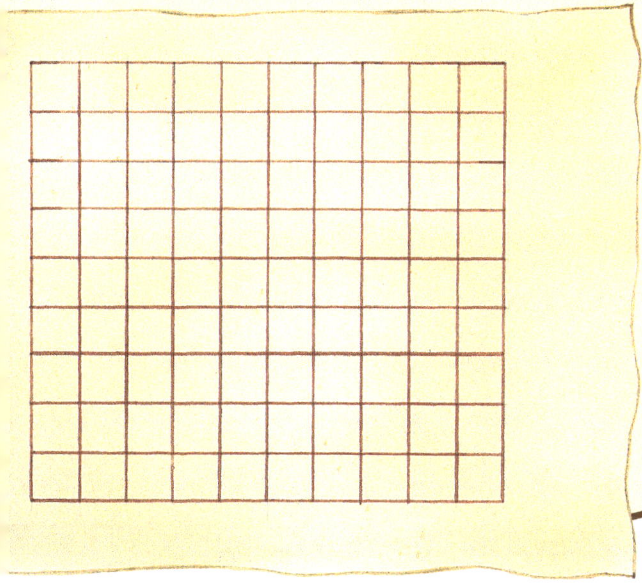

Of course, you could solve the problem much more quickly than Ahmose did. Just multiply the number of rows (9) by the number of blocks in a row (10) to get the answer: **9 x 10 = 90**.

For thousands of years, people didn't know about multiplication. When they wanted to know how many bricks were needed for a wall or how many tiles would cover a floor, they had to draw rows of squares, just as Ahmose did. Then they counted the squares. Sometimes the counting took hours!

About 500 years ago, someone made the first multiplication table. It is a chart that shows multiplication facts. To find the answer to a problem like 9 x 10, you look across the row that starts with 9 and down the column that starts with 10. Where the row and the column meet, you will find the answer!

Today most people memorize the multiplication facts instead of using a table. But you can check your answers with a multiplication table. Try a few problems and see!

3 x 7 = ?
8 x 5 = ?
4 x 6 = ?

Check your answers on page 182.

Multiplication Table

x	1	2	3	4	5	6	7	8	9	10
1	1	2	3	4	5	6	7	8	9	10
2	2	4	6	8	10	12	14	16	18	20
3	3	6	9	12	15	18	21	24	27	30
4	4	8	12	16	20	24	28	32	36	40
5	5	10	15	20	25	30	35	40	45	50
6	6	12	18	24	30	36	42	48	54	60
7	7	14	21	28	35	42	49	56	63	70
8	8	16	24	32	40	48	56	64	72	80
9	9	18	27	36	45	54	63	72	81	90
10	10	20	30	40	50	60	70	80	90	100

Multiplication Tricks and Tips

Here are some tricks that will help you learn to multiply quickly:

Any number multiplied by 0 equals 0.

2 x 0 = 0   0 x 2 = 0

9 x 0 = 0   0 x 9 = 0

Any number multiplied by 1 equals itself.

2 x 1 = 2   1 x 2 = 2

9 x 1 = 9   1 x 9 = 9

To multiply a number by 10, just put a 0 to the right of your original number.

2 x 10 = 20   10 x 2 = 20

9 x 10 = 90   10 x 9 = 90

Try This...

Here's a multiplication trick that will amaze your friends: Tell a friend to pick a number and not tell you what it is. Have your friend multiply the number by 2 and then by 5. Then ask for the answer. When you hear the answer, you will know what number she picked!

Here's how the trick works: The answer your friend says will end in a 0. Simply drop the 0 and you will know what number your friend picked. For example, suppose your friend picks the number 4. So, 4 x 2 = 8, and 8 x 5 = 40. When you hear the answer 40, drop the 0 and you'll know your friend picked the number 4!

Multiplying 9's is especially fun! You can use your fingers as a guide.

Choose a multiplication problem, such as 3 x 9. Hold up both hands, palms toward you. Then count three fingers (for three 9's), starting from your left thumb. Fold down the third finger on your left hand. You will see two fingers before the one you folded down and seven fingers after it. That's the answer to your problem:

3 x 9 = 27!

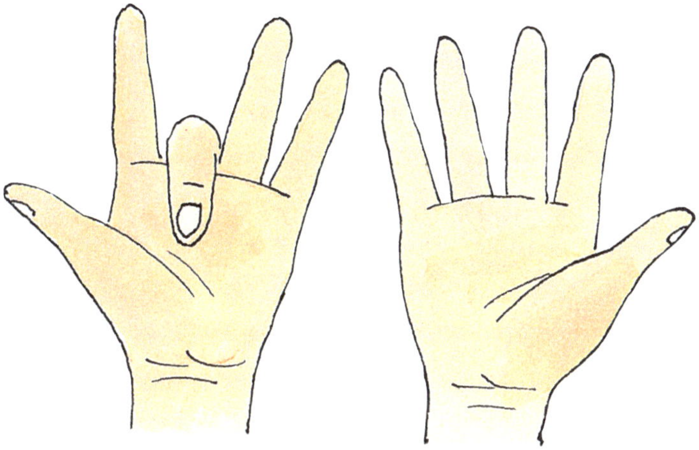

Now use your fingers to multiply 7 x 9. Count seven from your left thumb. Fold down your seventh finger, and there's the answer: **7 x 9 = 63!**

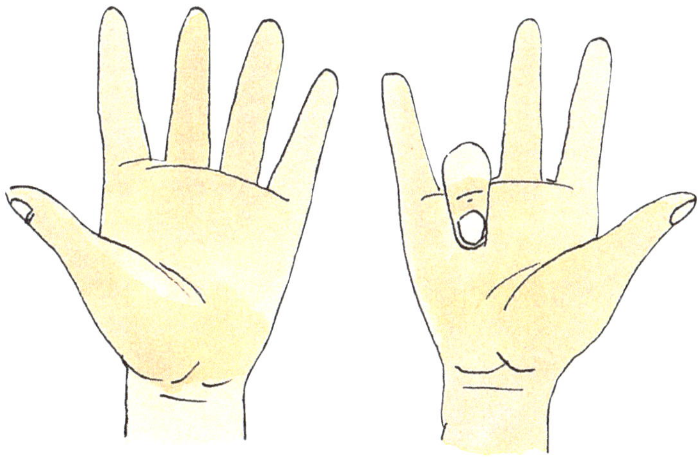

1x1 2x2 3x3 4x4

Square Numbers

You know what a square shape is. But did you know that a number can be square, too?

Four is a square number. It is the square of the number 2. You can use four candies or other markers to make a square: just put them in two rows of 2.

Three is not a square number. You cannot make a square out of three candies or other markers. Try it and see!

You can use multiplication to find square numbers. When you multiply any number by itself—for example, **1 x 1**, **2 x 2**, or **3 x 3**—the answer is a square number. Multiplying a number by itself is called **squaring** the number.

Below are pictures of the first six square numbers. How many candies would you need to make the next square number?

You can check your answer on page 182.

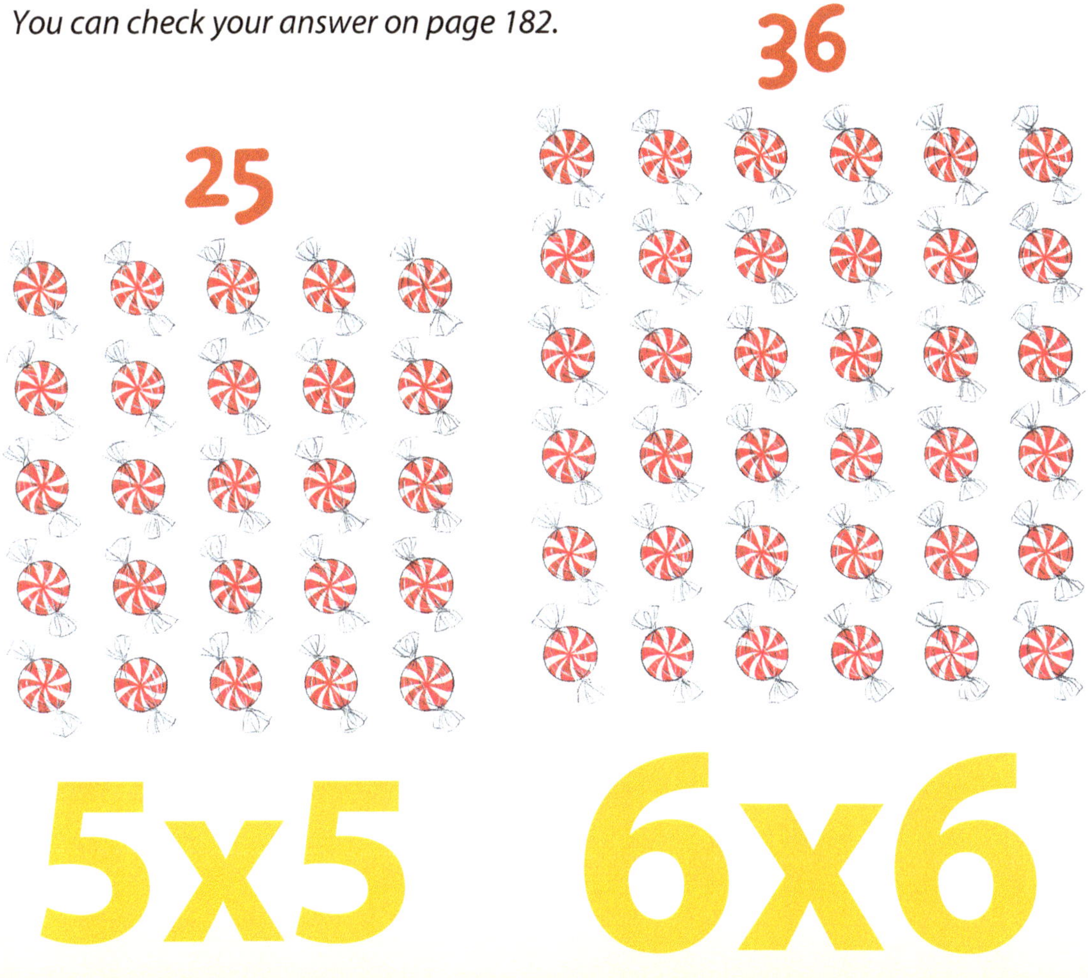

The Foolish Millionaire and the Clever Girl

Once there was a man who loved money more than anything else in the world. He was very rich, but he didn't know exactly how much money he had. One day he hired a little girl to count it for him.

The little girl counted for six days. Finally she came to the man and said, "You have 42 million dollars!"

The man was very pleased.

Slyly he asked the little girl how much she wanted to be paid for her work.

The girl said, "I worked for six days. Give me two pennies for the first day. For each day after that, give me the amount from the day before multiplied by itself."

The rich man thought about the girl's plan. For the second day, he would have to give her 2 pennies multiplied by 2 pennies, or 4 pennies. For the third day, he would give her 4 pennies times 4 pennies, or 16 pennies. For the fourth day, he would give her 16 pennies times 16 pennies, or 256 pennies.

The rich man stopped thinking and smiled. He would only have to pay the little girl a few dollars worth of pennies. What a foolish girl!

The rich man accepted the little girl's plan and began to count out her pay. She got 2 pennies for the first day, 4 pennies for the second day, 16 pennies for the third day, and 256 pennies for the fourth day of work. For the fifth day, she got 256 times 256 pennies, or 65,536 pennies.

For the sixth day, the little girl got 65,536 times 65,536 pennies, or more than 4 billion pennies—4,294,967,296, to be exact.

That equals more than 42 million dollars! The foolish millionaire had to give the little girl all his money. She was very clever to know about squaring numbers!

Using a Calculator

Did you check the math in the story about the foolish millionaire and the clever girl? You probably had no trouble multiplying **2 x 2** and **4 x 4** in your head. But what about **256 x 256**? It would take a while to solve that problem, even with a pencil and paper!

Many people use calculators to solve arithmetic problems involving big numbers. A calculator can solve difficult math problems in the blink of an eye—and it never makes a mistake!

Most calculators are easy to use. To enter a problem, you press a series of keys. An electronic system inside the calculator finds the answer. Almost immediately, the answer appears on a display window.

There are many kinds of calculators. Simple pocket calculators are useful for everyday tasks. Scientists, mathematicians, and engineers use more complicated models with dozens of keys.

This is a simple calculator. More complicated models may have dozens of keys.

Try This...

Calculator Games

Calculators are great for solving problems. They are also fun! Try these games with a simple calculator. If you haven't used a calculator before, ask an adult for help (and permission)!

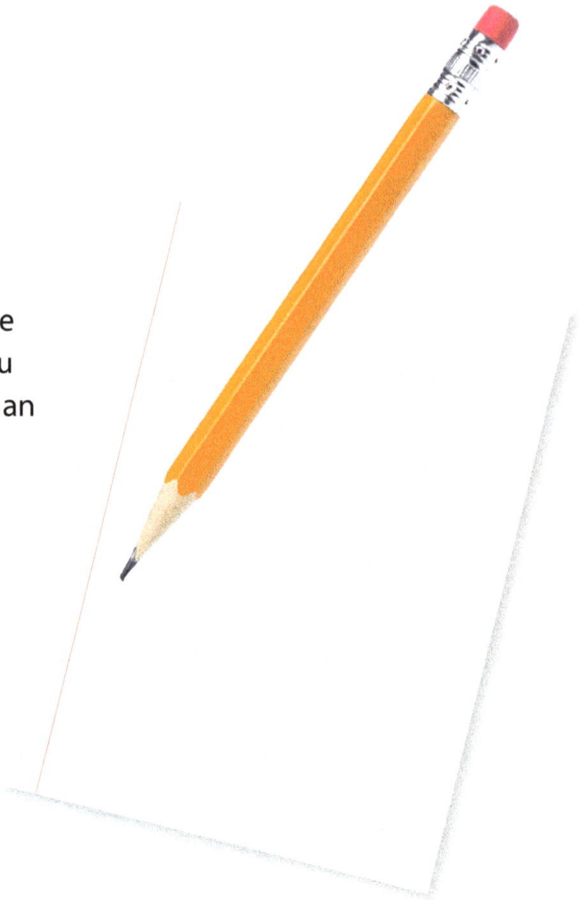

You Will Need:

- a calculator
- a pencil
- a piece of paper

Square Tricks

1. Think of a number. Enter it on the calculator.
2. Square your number (multiply it by itself) and write down the answer.
3. Press clear. Then enter your first number.
4. Add 1 and press the equal key.
5. Square that number.
6. Subtract the number you have written down.
7. Subtract 1.
8. Subtract your first number.

The answer will be your original number!

Note: If your calculator has a key marked x^2, you can press it to square a number.

Magic Nines

1. Write down a four-digit number (for example, 1,735). Don't use a number like 1,111 or 5,555 that has all four digits alike.

2. Mix up the digits to make a new number (for example, 7,513). Write down that number, too.

3. Which number is bigger? Enter it on your calculator.

4. Subtract the smaller number. Write down the answer.

5. Add up the digits in the answer. If you get a number with more than one digit, add those digits together.

No matter what number you start with, the answer will always be 9!

Texas Slim's Problem

Texas Slim and his group of bandits had just robbed a bank. They stole $35,000!

Slim grinned and rubbed his hands together. "Okay, boys," he said. "The seven of us will split this up fair and square."

"Hot dawg!" exclaimed Sagebrush Sam. "How much will that be for each of us?"

"I don't rightly know," said Slim. "I'll just start countin' it out and we'll see. Okay, here's a dollar for Sagebrush Sam, and one for Deadeye Pete, and one for the Sundown Kid, and one for . . ."

An hour later, Slim was still counting the money into seven piles. Then the sheriff and his deputies rode up and captured the whole group!

It's too bad Slim didn't know how to divide! He wanted to give each bandit an equal amount of money, so he needed to break $35,000 into seven equal parts. Slim tried to do this by counting out seven separate stacks. He put a bill on each bandit's stack, then put another bill on each bandit's stack, and so on. But that was a very slow way to count out each bandit's share.

Using division (duh VIHZH uhn), Slim could have figured out that each man would receive $5,000. Then he could have quickly counted out each bandit's money, and the group wouldn't have been captured. Of course, if Slim and his men were really smart, they wouldn't have stolen the money in the first place!

Division Delights

Imagine you are playing outside with three friends on a hot summer day. Your mother brings out a box of 12 popsicles. She says you may eat them all, but you must divide them evenly. How many popsicles will each of you get?

To solve this problem, you could keep subtracting four popsicles (one for each child) from 12 popsicles until nothing is left:

12 – 4 = 8   **8 – 4 = 4**   **4 – 4 = 0**

You can subtract four from 12 three times. This means that there are three 4's in 12. Each of you will get three popsicles!

Once you learn that, you know how to divide! You can write the problem like this:

12 ÷ 4 = 3 or $4\overline{)12}$ with **3** on top

We say, "Twelve divided by four equals three."

Basic statements in **division** are called division facts. You can calculate all the division facts by using subtraction. Here are two more division facts:

12 ÷ 6 = 2

12 ÷ 2 = 6

Do they look familiar? Compare them to these multiplication facts:

6 x 2 = 12

2 x 6 = 12

If you can multiply, you won't have trouble dividing because division is the opposite of multiplication!

More Number Fun

When you toss a coin,
which side do you call?
Is "heads" or is "tails"
more likely to fall?

What's in the message?
Do you know the code?
Can you figure the meaning
before you've been told?

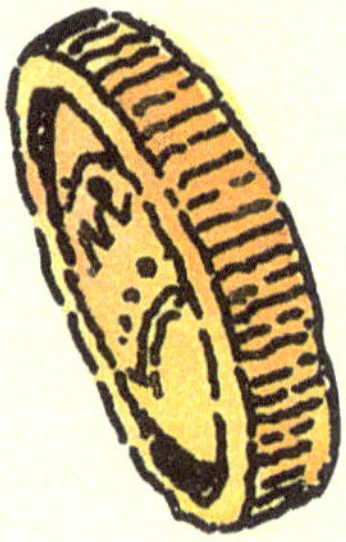

When you play a game,
do you tally your score?
You can use numbers
for this and much more.

Heads or Tails

Rick wanted to go to the park. Gayle wanted to go swimming. They agreed to toss a coin to decide what they would do.

"Heads or tails?" asked Rick as he flipped the coin.

"Tails never fails!" called Gayle. Sure enough, the coin landed with the tails side up.

"See?" Gayle grinned. "I told you that tails never fails! Swimming it is."

Do you think Gayle is right? Does tails win more often than heads? Try this experiment and find out.

Try This...

Tally marks are an easy way to keep track of numbers when you're counting. Use four straight lines to represent the first four things. Then draw another diagonal line through the four lines to represent the fifth thing. Start with another straight line for the sixth thing. When you're finished, count the groups of five and add the few remaining lines as ones.

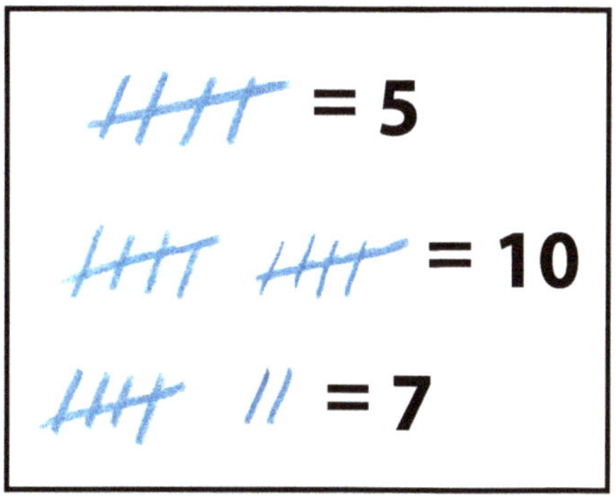

Make two columns on a piece of paper. Label them "Heads" and "Tails." Flip a coin 100 times. On your paper, mark which side shows after each flip. Use tally marks in each column.

Count the marks in each column. There may be a few more heads or a few more tails, but the numbers should be almost the same.

Every time you flip a coin, the chance that it will come up heads is exactly the same as the chance that it will come up tails. We say that the chances are even.

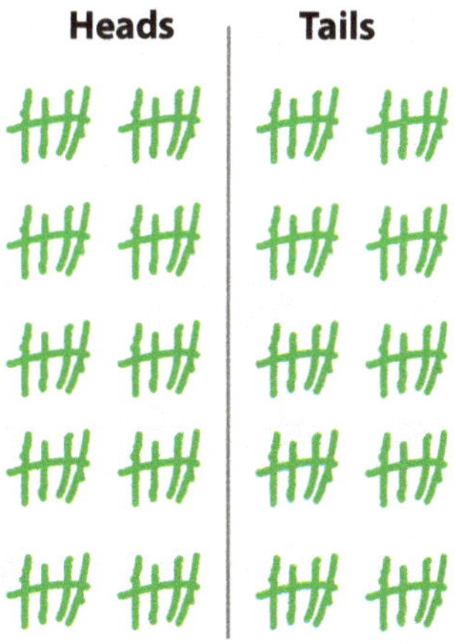

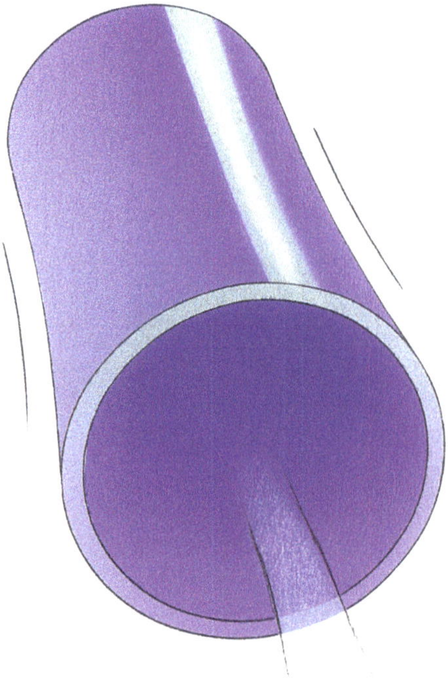

The Luck of the Dice

When you flip a coin, the chances of getting heads are the same as the chances of getting tails. Rolling a die is similar, but a die has six sides. The chances of rolling a 1 are the same as the chances of rolling a 2, 3, 4, 5, or 6.

Everything changes when you roll two or more dice. Imagine that you and a friend are playing a game with two dice. Pretend that if you roll a 12, you'll win the game, but if your friend rolls a 7, she'll win.

Which of you is more likely to win? Or are your chances about the same?

When you roll two dice and add their numbers together, you get a number between 2 and 12. There are 36 possible combinations of two dice.

There is only one way to get a 12. You have to roll two 6's. So you have one chance out of 36 (1/36) to roll a 12.

But there are six ways to roll a 7! So your friend has six chances out of 36 (6/36) to roll a 7. That means her chance of winning the game is better than yours!

Try This...

Roll a pair of dice 50 times. Keep track of the numbers that come up. Which numbers come up most often?

Check your results with the answers on page 183.

There's only one way to roll a 2 or a 12.

There are two ways of rolling a 3 or an 11.

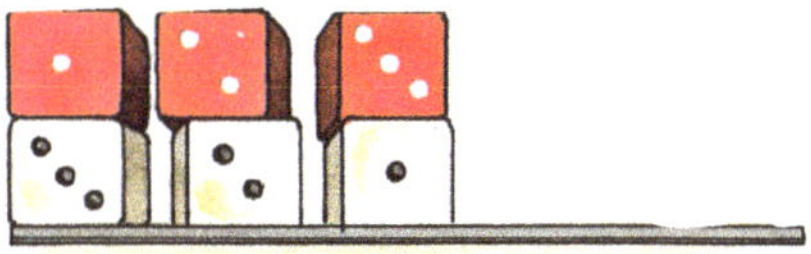

There are three ways of rolling a 4 or a 10.

There are four ways of rolling a 5 or a 9.

There are five ways of rolling a 6 or an 8.

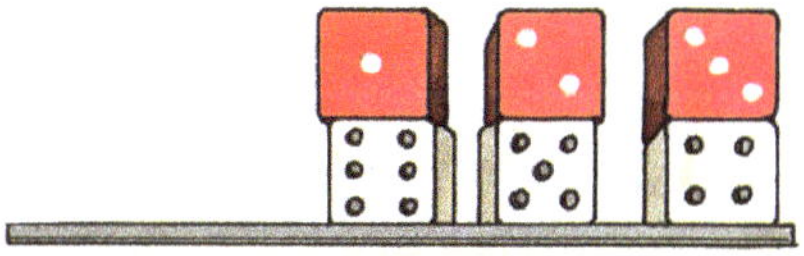

There are six ways to roll a 7.

Sharing Birthdays

What would you do if you were invited to two birthday parties on the same day?

You would probably go to the party of the person who asked you first. That would be the fair thing to do.

It's not unusual to hear about two parties on the same day. But do you know two people who actually have the same birthday?

Let's say you have 30 friends. With 365 days in a year (ignoring leap days that occur every four years), what are the chances that two of your friends will have the same birthday? Pretty slim, you might think.

But you'd be wrong! With billions of people in the world and only 365 days on which to have birthdays, millions of people share the same birthday. In any large group of people, it's likely that two or more will have the same birthday.

In fact, in a group of 30 people, it's more likely that at least two people will have the same birthday than that all 30 people will have different birthdays!

Try This...

Ask 10 people when their birthdays are. Make a list. All 10 people will probably have different birthdays.

Then ask 10 more people when their birthdays are. Add these dates to your list. Out of 20 people, you may find two people who have the same birthday.

Finally, ask 10 more people when their birthdays are. Out of 30 people, you'll probably find two or more who share a birthday!

Ali Kwazoor's Choice

Prince Ali Kwazoor of Zuristan had sailed stormy seas and crossed burning deserts. He had fought fierce dragons and scary monsters. But now his journey was nearly over.

Prince Ali had reached the Mountain of Darkness. There, in a cave, was the thing he had traveled so long and far to find—the great Treasure of Samarkand!

As Prince Ali entered the cave, there was a flash of lightning and a clap of thunder. An old man in a flowing robe suddenly appeared.

"I am the Wizard of Kandev," he said. "I guard the Treasure of Samarkand. You must pass a test before you win the great treasure."

The wizard held out two boxes. One box was red; the other was yellow. The red box contained four pebbles— one black pebble and three white ones. The yellow box contained seven pebbles— three black ones and four white ones.

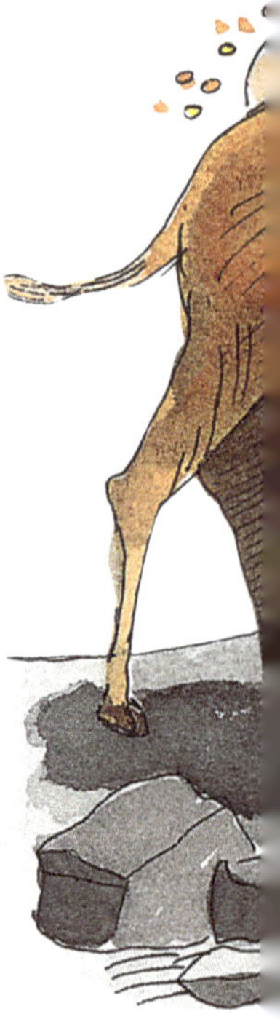

"Here is your test," said the wizard. "Without looking, you must take a pebble from one of the boxes. If you pick a black pebble, the treasure is yours. But if you pick a white pebble, you will be turned to stone!"

Ali thought carefully. Should he pick a pebble from the red box or the yellow box?

In the red box, there were four pebbles, one of them black. That gave Ali one chance out of four to pick a black pebble. In the yellow box, there were seven pebbles, three of them black. That gave him three chances out of seven to pick a black pebble.

Ali closed his eyes and picked a pebble from the yellow box. Slowly he opened his eyes and looked down at the pebble. It was black! Prince Ali went home with the great Treasure of Samarkand.

Try This...

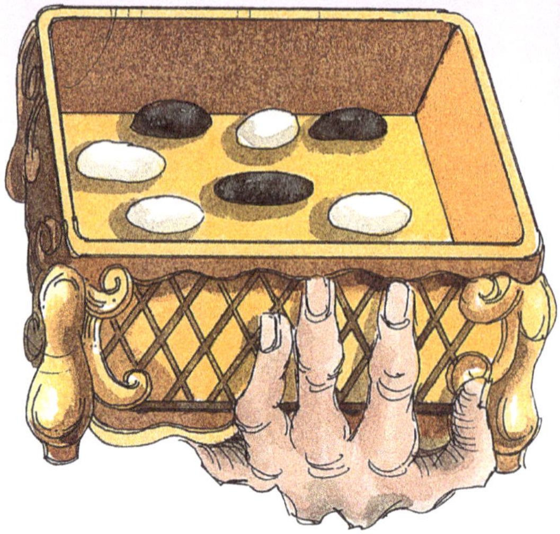

Take the Wizard's Test

Was Prince Ali wise to choose a pebble from the yellow box? Which box would you have picked from? Try this activity to find out which box offers a better chance of picking a black pebble.

You Will Need:

- 4 red checkers
- 3 black checkers
- a bag or box
- paper
- pencil

What To Do:

1. Place the four red checkers and three black checkers in the bag or box.
2. Make two columns on your paper. Label them "Red" and "Black."
3. Without peeking, put your hand in the bag or box and pull out a checker. What color is it? Make a tally mark in the correct column on your paper. Then put the checker back and shake the bag or box.

4. Do this 49 more times. Use tally marks in each column to record how many times you picked red and black checkers.
5. Remove one red checker and two black checkers from the bag or box. Start new columns for "Red" and "Black" on the other side of your paper. Then repeat steps 3 and 4 with three red checkers and one black checker in the bag or box.
6. Compare your results. Did you pick a black checker more often in the first experiment or the second experiment? Most likely, your results show what Prince Ali knew—three out of seven (3/7) is a better chance than one out of four (1/4). But Ali was still fortunate to pick a black pebble, wasn't he?

Try This...

Buttons and Boxes

Here's a puzzle for you to solve! How many ways can you arrange five different buttons in two boxes without repeating any of the ways you arrange them?

You Will Need:

- 2 small boxes
- 5 different-colored buttons
- 5 crayons in the same colors as the buttons
- paper

What To Do:

1. Put one button in one box and the other buttons in the other box. We call that a combination (KAHM buh NAY shuhn) of the buttons.

2. With your crayons, draw a picture of the combination of the different colored buttons.
3. Move a button from the second box to the first box.
4. Draw a picture of the new combination.
5. Continue arranging the buttons and drawing pictures of the combinations. Use the color of the buttons to help you make different combinations.
6. How many different combinations can you make with five buttons and two boxes? Make sure you have not repeated any combinations. Then compare your pictures with the pictures shown below. The buttons are numbered to help check your answers.

As you see here, there are 16 possible combinations. Did you find them all?

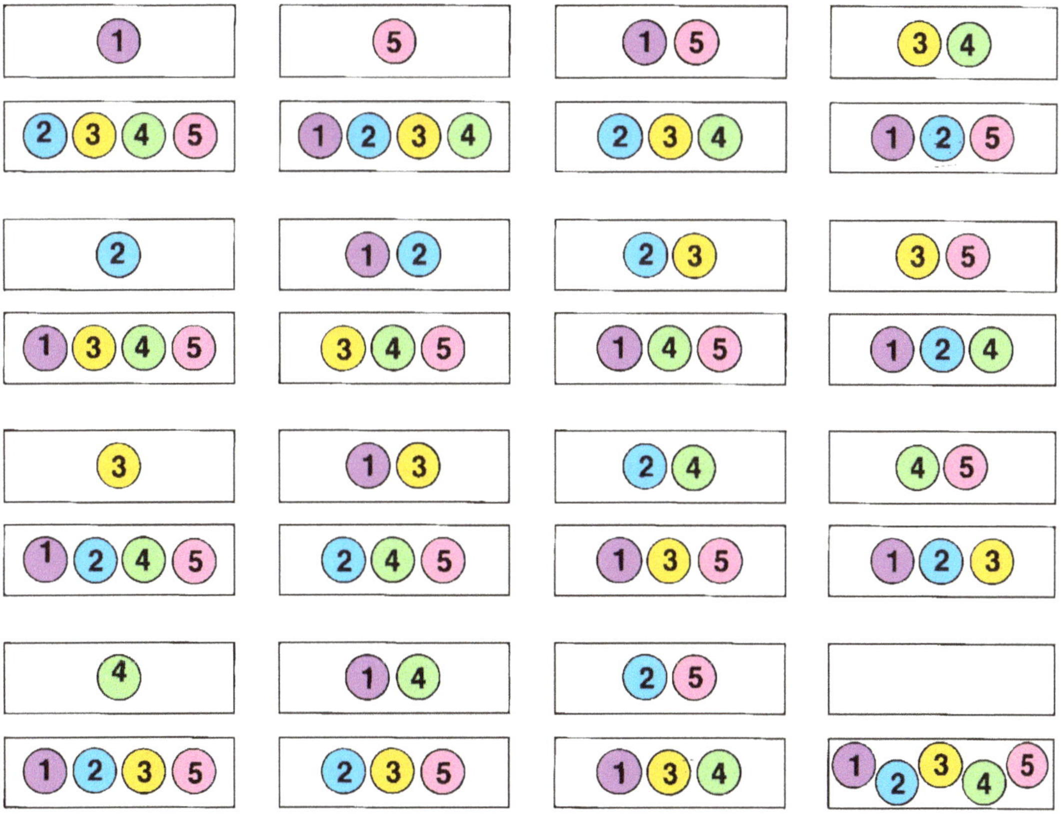

Holmes and the e

Have you heard of Sherlock Holmes? He is a detective character created by the British writer Sir Arthur Conan Doyle. Holmes is the main character in dozens of mystery stories.

In one story, called "Adventure of the Dancing Men," Sherlock Holmes solves a mystery by figuring out a message written in code. Holmes knows that the letter *e* is used more often than any other

Try This...

Write the letters of the alphabet on a piece of paper. Then turn to any page in this book—or any book. Count the first 100 words on the page and mark where they end with a slip of paper. Then use tally marks to record how many times each letter of the alphabet appears in those words. Which letter was used most often? Usually, people have more marks for the letter e than for any other letter.

letter in the English language. So he guesses that the symbol appearing most often in the coded message might stand for *e*. From that beginning, he is able to figure out the rest of the message.

How does Sherlock Holmes know that e is the most-used letter in English? Try the experiment at the top of the page and find out!

Fun with Codes

Did you ever want to send someone a secret message—a message that only you and that person would understand?

One way to write a secret message is by using a code. A code is a system of letters, numbers, or other symbols that stand for letters or words in a message.

Did you read "The Minus Mystery" on pages 104-105? In that story, Mildred Minuette wrote a secret message to Penelope Puzzle. Her code used numbers. Each number stood for a letter: A = 1, B = 2, C = 3, and so on.

Below is a message in a different code. Again, each number stands for a letter. See if you can figure out the code and read the message. If you need help, read the hint below.

8 22 24 9 22 7 14 22 8 8 26 20 22

Hint: 2 = Y.

Try This...

Send your own secret message to a friend!
Use della Porta's code or make up your own.

The code to the right is more difficult. It was invented by Gianbattista della Porta, an Italian who lived in the 1500's and early 1600's.

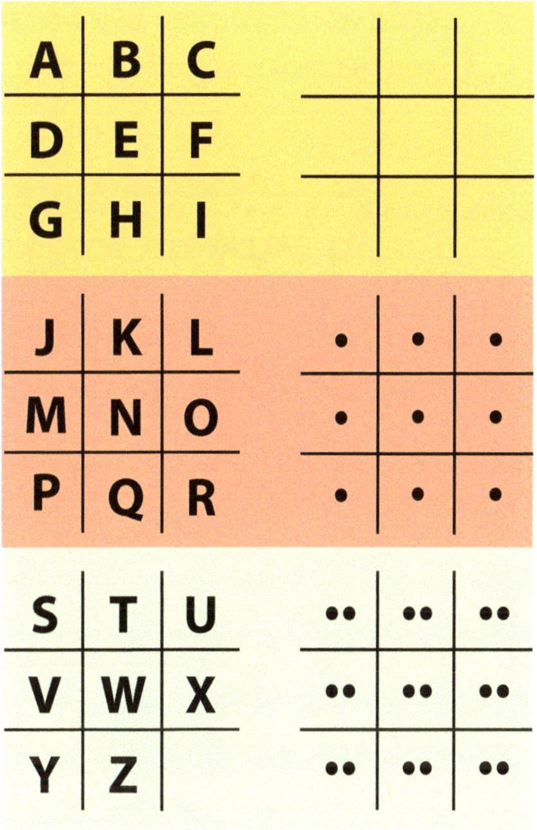

Can you read this message in della Porta's code?

Now write this message using della Porta's code:

I LIKE CODES

Check your answers on page 183.

More Fun with Codes

Here are more secret messages! See if you can decode them—undo the code and read the messages. Some codes use numerals, and others use letters. Once you figure out how a code works, you can use it to send messages to your friends! If you get stuck, you can find hints at the bottom of this page.

1. **GSV ZOKSZYVG RH GFIMVW ZILFMW**
2. **ITI SALLIN HO WYO ULOO KATIT**
3. **2 11 19 23 2 21 14 3   21 7   20 25 14**
4. **HKZ XVWW RUUV IVMGO BZMWYZ XPDZIW**

1. **Hint:** Think backward!
2. **Hint:** Look at this message carefully!
3. **Hint:** A = 1, D = 2, G = 3
4. **Hint:** This code combines two of the other codes. Decode the message once, then look at it carefully.

5. Imagine that you found a rolled-up paper in a bottle on the beach. The paper has a message: Found pirate treasure! You won't believe your eyes! Come to

VJG UEKGPEG OWUGWO.

Can you decode the message and find the pirate treasure? There is a hint below if you need one.

Check your answers on page 183.

5. **Hint:** In this code, today = vqfca.

Try This...

Statistics and Sports

Statistics (stuh TIHS tihks) are numbers that give us information. For example, they tell how many people live in your city—and how many of them own cars or televisions. They tell the average (usual) temperature for your city on any date.

Statistics are an important part of sports, too! They tell how many games your favorite team has won and lost. They tell which soccer player has made the most goals this season. They tell which basketball player scores the most points per game.

Joel, Ahmed, and Fernando play on a soccer team called the Rockets. The Rockets played four games this month. See how many goals Joel, Ahmed, and Fernando got in each game:

Game	1	2	3	4
Joel	4	1	2	0
Ahmed	1	0	1	2
Fernando	2	3	1	2

Use the chart above to answer these questions:

1. Which player made the most goals in one game?
2. Which player made the most total goals?

Some statistics for a basketball team called the Stars are on page 153. See how many points each player scored in the last game.

Player	Points Scored
Carmen	10
Josie	3
Alexis	7
Maya	2
Nujanart	0
Debby	4
Kirsten	8
Tanika	4
Patricia	0
Chris	0
Elena	2
Jean	0

Use the chart above to answer these questions:

3. How many points did the team score?
4. Which player got the most points?
5. How many players are on the team?

You can check your answers on page 183.

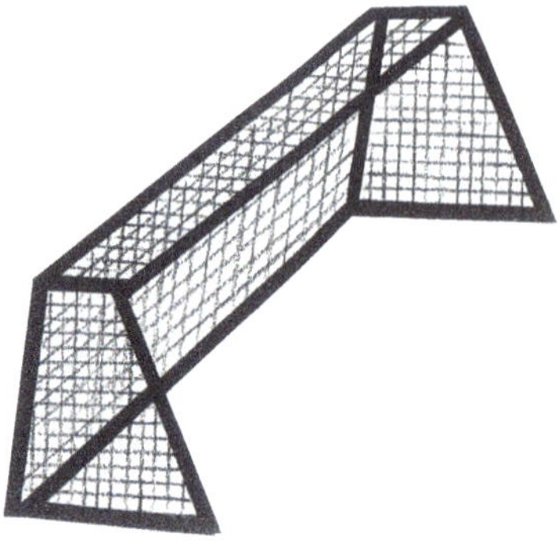

What Is a Graph?

A **graph** (GRAF) is a picture. Graphs show statistics in a way that is easy to understand. There are many kinds of graphs.

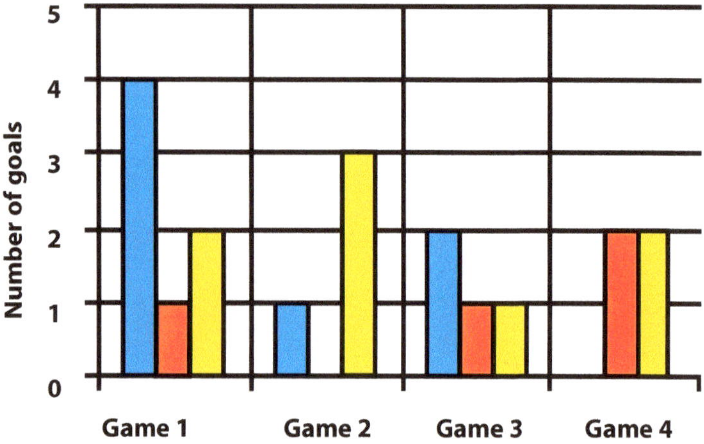

The graph above is a bar graph. It shows how many goals Joel, Ahmed, and Fernando made in four soccer games.

The numbers on the left are the number of goals. Joel's goals are shown in blue, Ahmed's goals are in red, and Fernando's goals are in yellow. Notice that Ahmed made 0 goals in Game 2. Who made 0 goals in Game 4?

Who made the most goals in one game? You can tell very quickly by looking at the bar graph.

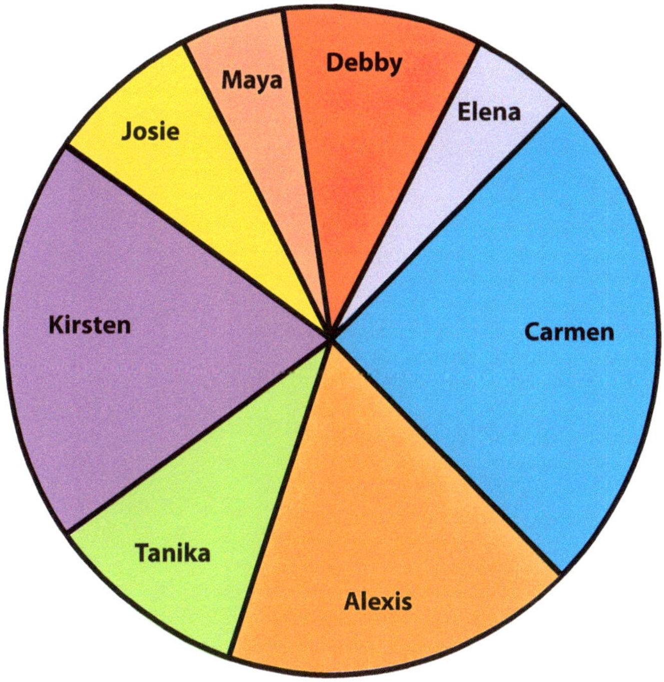

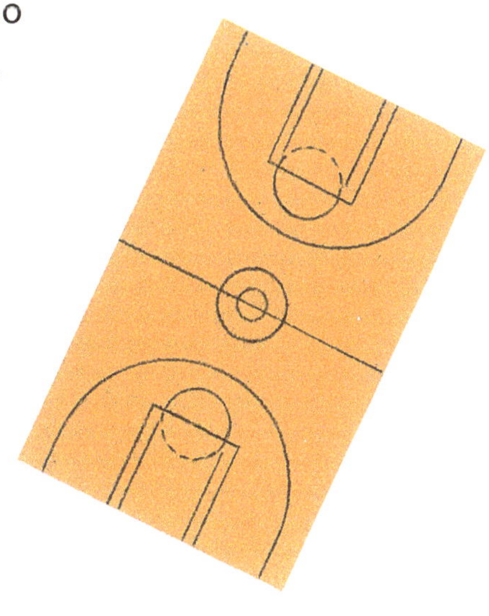

Above is a circle graph or "pie chart." It shows the points scored by members of the Stars basketball team during a game.

The circle stands for the total number of points scored by the team (40). Each colored section stands for the points scored by a particular player. There are no sections for players who scored 0 points.

Which section is the biggest? What does that tell us?

Check your answers on page 183.

Ways of Measuring

Feet and meters, pounds and grams,
hours and months and days.
How did we come to measure things
so many different ways?

It all goes back to long ago,
in many ancient lands,
when people started measuring
with fingers, feet, and hands.

They measured time by shadows
and with dripping water clocks.
They measured years by seasons.
They measured weight with rocks.

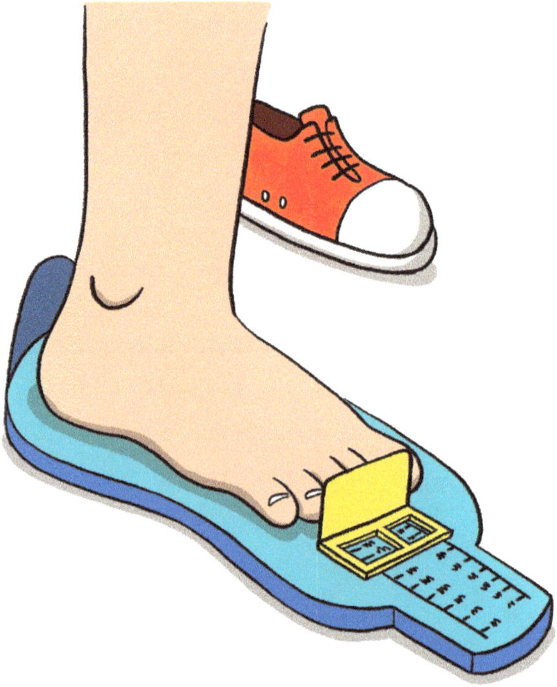

What Is Measuring?

Measuring means finding the size of something. Think of all the things you can measure! You can measure how tall you are and how much you weigh. You can measure the size of your computer. You can measure wrapping paper to fit around a present. You can measure ingredients to use in a recipe.

Do we measure all these things the same way? No. There are many ways to measure.

When you measure how tall you are, or how long your foot is, you are measuring length.

When you measure how heavy you are—or how heavy a bunch of bananas is—you are measuring **weight**.

When you find the size of a flat shape like a piece of paper or a football field, you are measuring **area** (AIR ee uh).

When you measure how much cereal will fit in a bowl or how much water your bathtub will hold, you are measuring **volume** (VAHL yuhm).

How do we measure time? We use a clock to measure seconds, minutes, and hours. We use a calendar to measure days, months, and years.

KNOW IT ALL

It's often helpful to guess the size of something when you can't measure it. We call this kind of guessing **estimating** (EHS tuh mayt ihng).

Beginning to Measure

Long ago, people didn't have rulers or scales to help them measure things. So they used what they had—their arms, hands, and feet! You can measure this way, too. Try measuring the length of your bedroom with your feet. How many "feet" long is it? Then ask a grown-up to measure your bedroom the same way. Do the answers match?

As you probably discovered, some people have longer arms, wider hands, or bigger feet than other people. So measurements using body parts don't always match. This can lead to big problems and wrong measurements!

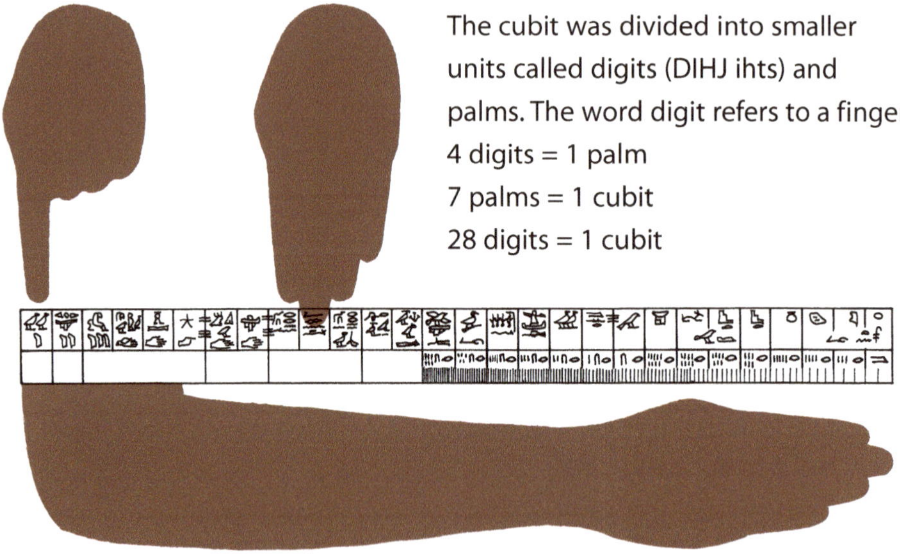

The cubit was divided into smaller units called digits (DIHJ ihts) and palms. The word digit refers to a finger.
4 digits = 1 palm
7 palms = 1 cubit
28 digits = 1 cubit

Try This...

Use a tape measure to find the length of a friend's arm from the elbow to the tip of the middle finger. Then ask your friend to measure your arm the same way. Do the lengths match? How close are they to an Egyptian cubit? Measure some other people's arms. Make a graph to show the lengths.

The ancient Egyptians may have been the first people to create standard measurements. That means measurements that are the same all the time. They had a unit of length called a cubit (KYOO biht). A cubit was about 21 inches (53 centimeters) long.

When they invented the cubit, the Egyptians measured someone's arm—probably the pharaoh's—from the elbow to the tip of the middle finger. Then they made a bar of black granite that was exactly this length. This bar became the official cubit. People no longer measured with their own arms. Instead, they used cubit sticks whose length matched the official cubit bar.

This picture shows part of a real Egyptian cubit stick. It is about 3,500 years old!

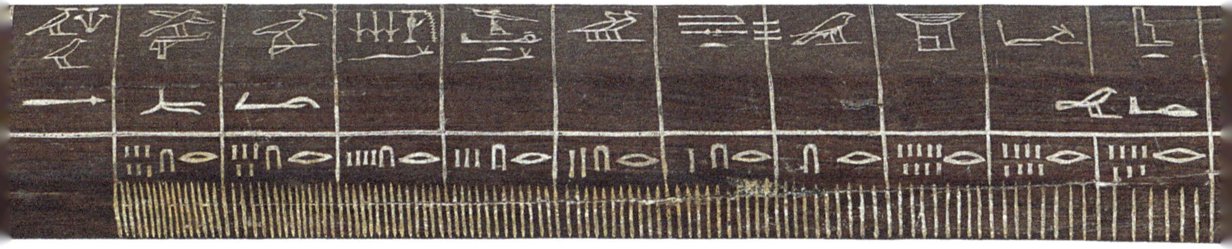

Units of Measurement

Numbers are an important part of measuring. But they're not the only part! When someone asks how tall you are, you don't say "52" or "132." If you're in the United States, you say, "52 inches." If you're in China, you say, "132 centimeters." Inches and centimeters are units of measurement.

Inches are part of the **inch-pound** system of measurement commonly used in the United States. One inch is as long as this line:

Other units of length in the inch-pound system are called feet, yards, and miles. There are 12 inches in 1 foot. There are 3 feet in 1 yard. A mile is much longer. There are 5,280 feet in 1 mile.

In the late 1700's, the French government asked scientists to invent a new system of weights and measures. The scientists wanted the system to be simple and based on facts in science. That's how the **metric system** was created. The metric system is based on 10's. It is used in most countries of the world. The United States uses metric measurements for science, medicine, and some sports.

Centimeters are part of the metric system. One centimeter is the length of this line:

Other units of length in the metric system are meters and kilometers. There are 100 centimeters in 1 meter. There are 1,000 meters in 1 kilometer.

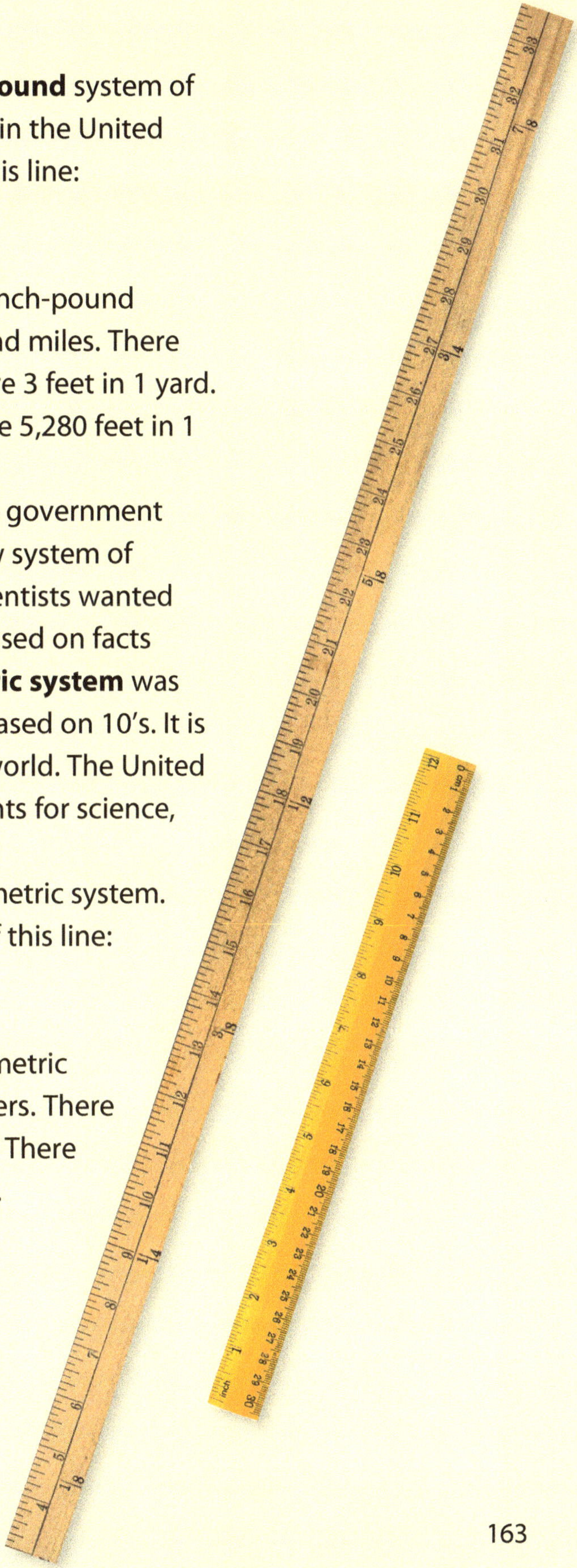

Try This...

How Long Is It?

How long is your foot? How about your hand? You can measure the length of your own hand and foot as well as many objects around your home.

You Will Need:

- objects to measure (see activity directions)
- paper clips
- playing cards
- paper
- a pencil

What To Do:

1. Make a line of paper clips to measure each of the objects that follow. Don't leave any space between the paper clips. Place them together end-to-end. Draw a bar graph to show the lengths.

   a. a pencil

   b. a toothbrush

   c. a spoon

Round off your measurements to whole paper clips, as shown here.

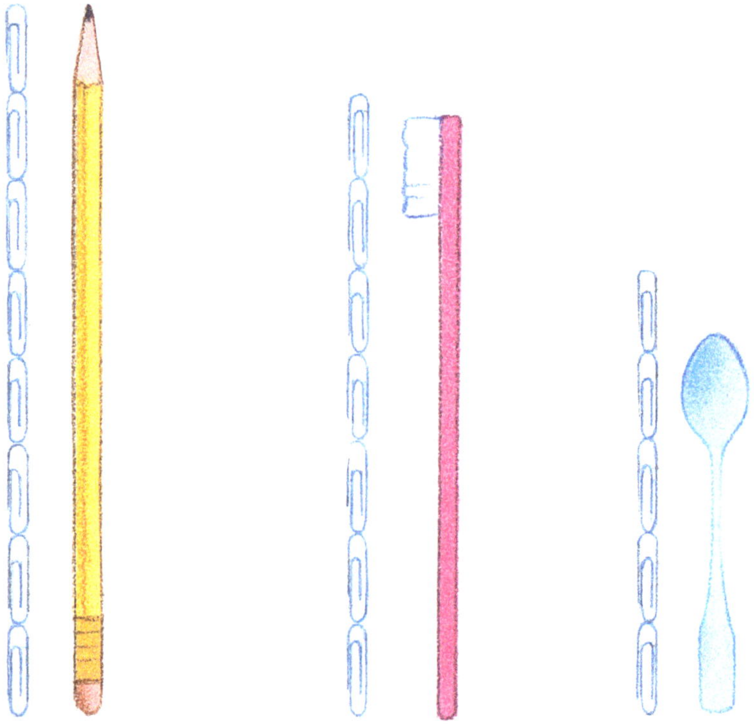

8 paper clips **7 paper clips** **4 paper clips**

2. Estimate the lengths of the following objects in paper clips. Then measure them. Write the estimated lengths and actual lengths on a piece of paper. How close were your estimates?

   a. a marker

   b. a fork

   c. a crayon

   d. your hand, from wrist to fingertip

   e. your foot, from heel to tip of longest toe

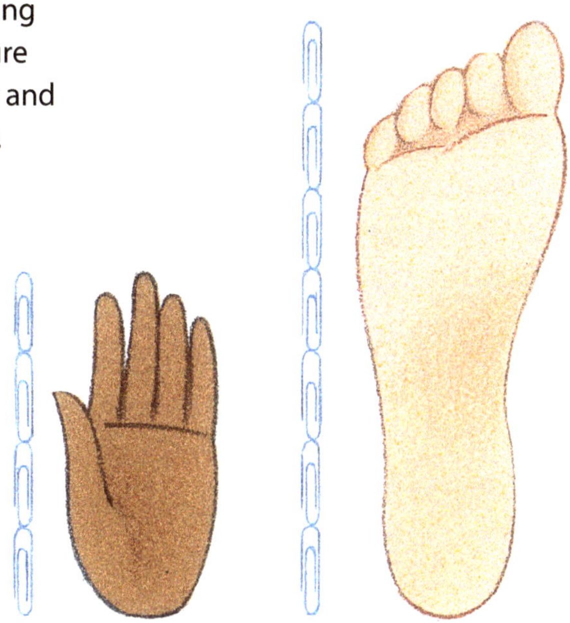

4 paper clips **7 paper clips**

3. Use playing cards to measure the lengths of the following objects. Make a bar graph to show the lengths.

   a. a magazine

   b. a baseball bat

   c. your bed

Try This...

How Much Does It Weigh?

How well can you estimate the weights of things? Try these activities and find out!

You Will Need:

- a bathroom scale
- 5 grocery bags
- an assortment of heavy objects, such as books or rocks
- a pencil
- paper

Use a Bathroom Scale:

1. Put several objects in each grocery bag. Lift the bags to see how heavy they feel.
2. Number the bags in order, from the one you think is heaviest to the one you think is lightest.
3. Weigh each bag on the scale. Did you put the bags in the correct order?
4. Draw a bar graph to show the weights of the bags.

You Will Need:

- 2 manila envelopes
- a hole punch
- string
- scissors
- a hanger
- a clothesline
- tape
- small objects to weigh (for example, crayons, checkers, feathers)

Use a Balance Scale:

1. Punch a hole in the flap of each envelope. Thread string through the holes, and tie one envelope to each end of the hanger.
2. Hang the hanger on the clothesline. Adjust the envelopes so the hanger hangs evenly. Tape the strings to the hanger.
3. Pick up two different objects. Try to guess which one is heavier. Test your guess by putting one object in each envelope. The hanger will tip toward the heavier object.
4. Try several pairs of objects. The more you practice, the better you'll get at estimating which object is heavier!

Try This...

What Is the Area?

Which of these shapes on the next page is the biggest?

To find out, you can measure their area—the space inside each shape.

You Will Need:

- white paper
- a pencil
- glue
- lightweight cardboard
- scissors

What To Do:

1. Trace the squares below on white paper.
2. Glue the sheet of squares onto the cardboard.
3. Cut out the individual squares.

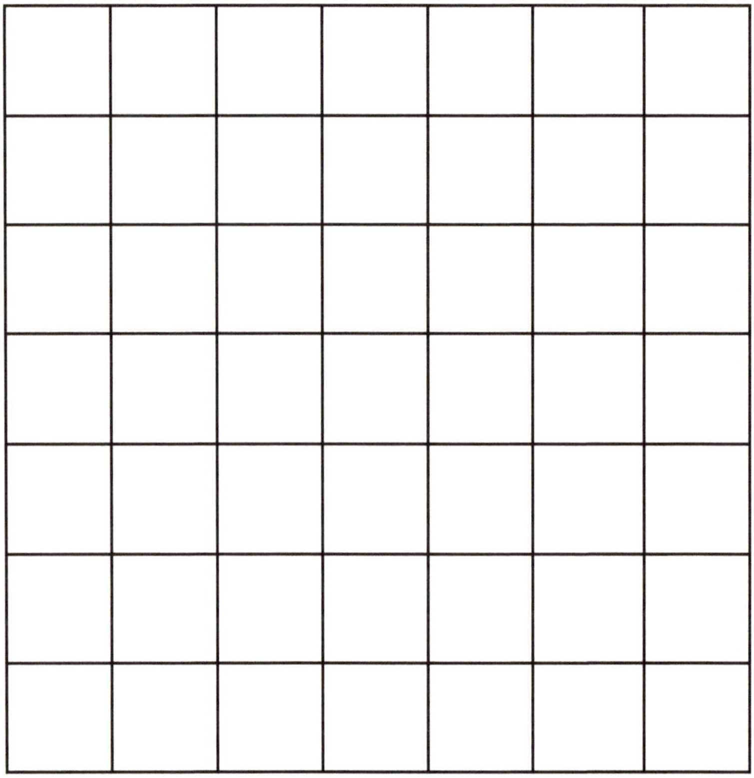

4. Lay the squares inside the castle shape above. Don't leave any space between the squares. How many squares fit inside the castle?

5. Now lay the squares inside the skyscraper shape to the left. How many squares fit inside it?

6. Compare how many squares you put in the castle with how many you put in the skyscraper. Which number is bigger? That tells you which shape has the bigger area!

Check your answers on page 183.

Try This...

How Much Will It Hold?

Do you like to cook? Measurement is an important part of cooking. Some recipes call for such ingredients as 2 cups of flour or 1 teaspoon of lemon juice.

Cups and teaspoons are two units for measuring volume. The volume of a container is the amount it can hold. You can measure volume without special measuring cups or measuring spoons. Try this activity and find out!

You Will Need:

- a pitcher of water
- a large pan of uncooked rice or dried beans
- several sizes of plastic drinking cups, bowls, and other containers
- a small scoop or spoon

What To Do:

1. Choose a drinking cup and a bowl. Estimate how many of those cups full of water will fit in the bowl.
2. Fill the drinking cup with water and pour it into the bowl. Continue until the bowl is full. How many of these drinking cups of water does it take to fill the bowl? How close was your estimate?

3. Repeat steps 1 and 2 with other sizes of drinking cups and bowls.
4. Now, choose two different-shaped containers. Which one do you think has the bigger volume?
5. Use the scoop to fill both containers with rice or beans. Count the number of scoops you put in each container. Which container holds more? Was your estimate correct?
6. Repeat steps 4 and 5 with other pairs of containers.

Note: Do not drink the water or eat the foods after you are done with this experiment.

Try This...

Checking Measurements

Try these measuring experiments!

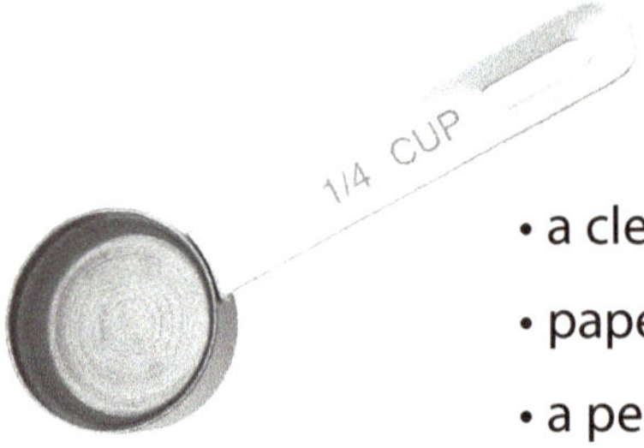

You Will Need:

- chalk
- a yardstick
- a meterstick
- a ¼-cup measuring cup
- small candies or pieces of cereal
- a clear plastic jar with a lid
- paper
- a pencil
- a scale

What To Do:

1. Ask permission to draw with chalk on a sidewalk or driveway. Estimate and draw the following lengths. Then use the yardstick and meterstick to measure to check your estimates.

   a. 5 inches

   b. 1 meter

   c. 2 yards

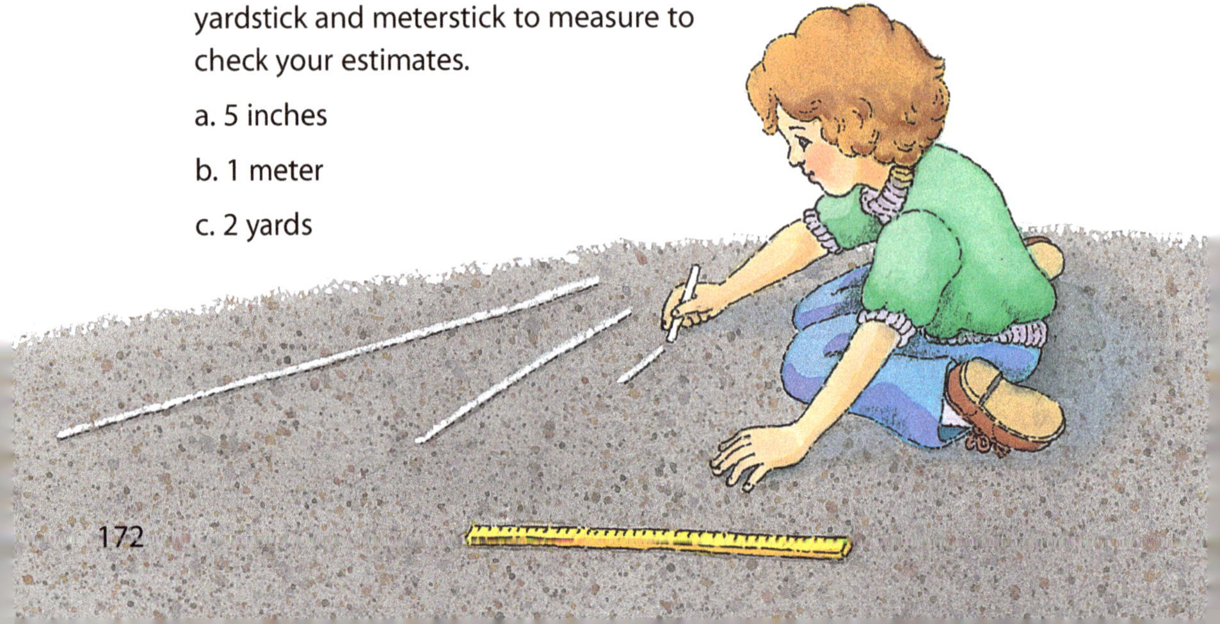

2. Measure ¼ cup of small candies or pieces of cereal. Count the pieces. Then measure 2 cups of the same candies or cereal into a jar. Multiply the number of pieces in the ¼ cup by 8 to estimate how many pieces are in the jar. Then ask several friends to estimate how many pieces are inside. Whose estimate comes closest to yours? If you like, count the pieces to find out exactly how many pieces are in the jar.

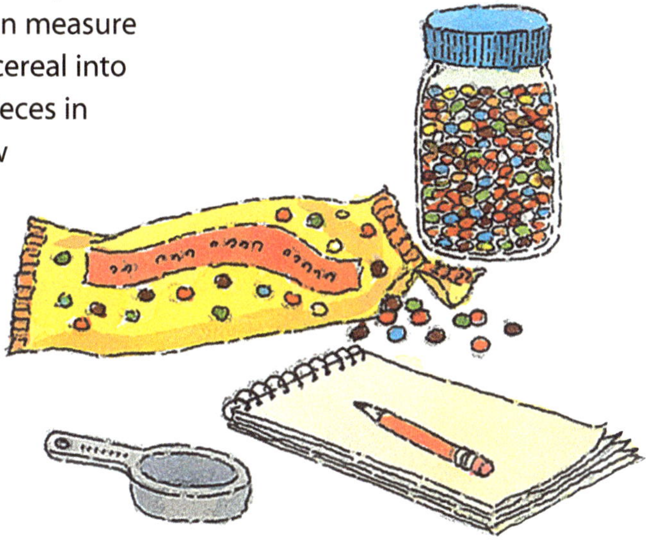

3. At home, weigh various items that you find on a scale. You can also weigh yourself! First estimate how much each item will weigh. How close are your estimates?

What Time Is It?

Here's a riddle: How is a clock like a ruler?

Answer: They both measure things! A ruler measures length. A clock measures time.

For thousands of years, people have used shadows to measure time. In the morning and evening, the sun is low and shadows are long. At midday, when the sun is high, shadows are short or may even disappear.

More than 4,000 years ago, people began to use sundials. A sundial is a clock that uses shadows to tell time. Sundials work very well, but they can only be used outside or near a window when the sun is shining!

Try This...

You can make a sundial with a paper plate and a pencil. Take the plate outside to a sunny, flat space. Poke a pencil through the center of the plate. Push the pencil into the ground. Then push the plate down until it is flat on the grass or dirt. Be careful not to move or spin the plate during your experiment. Every hour on the hour, draw a line to mark the shadow of the pencil on the plate. Write the time below it. Put your sundial outside in the same position on sunny days and use it to tell time!

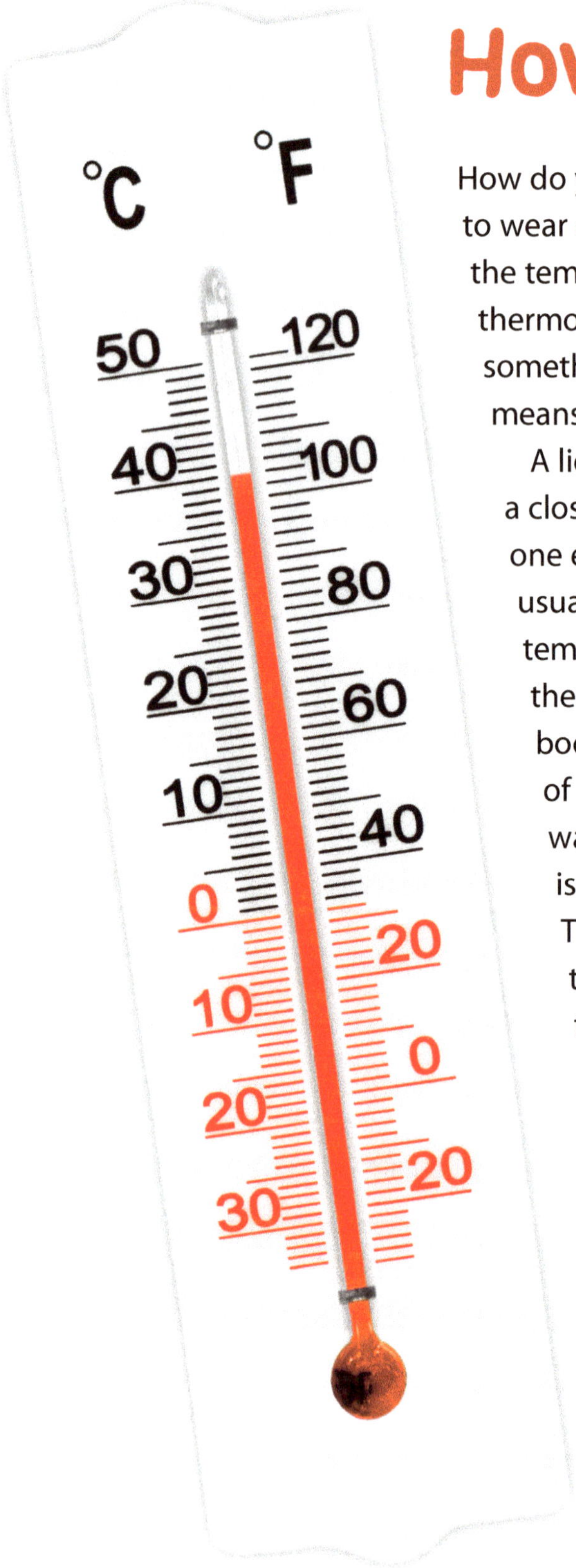

How Hot Is It?

How do you know what kind of clothes to wear each day? One way is to check the temperature on a thermometer. A thermometer measures how hot or cold something is. The word *thermometer* means "heat measurer."

A liquid-in-glass thermometer is a closed glass tube with a bulb at one end. These thermometers are usually used to measure outdoor temperatures. Different kinds of thermometers are used to take your body's temperature. The bulb is full of liquid. When the liquid becomes warm, it moves up the tube. When it is cool, it falls back toward the bulb. There are numbers and lines next to the tube which indicate the temperature at every level.

If you don't know how to read a liquid-in-glass thermometer, ask an adult to show you. Then try the activity on the next page.

Try This...

Measure the Weather

Do you think there are patterns to the weather? This experiment will help you find that out and practice measuring the temperature.

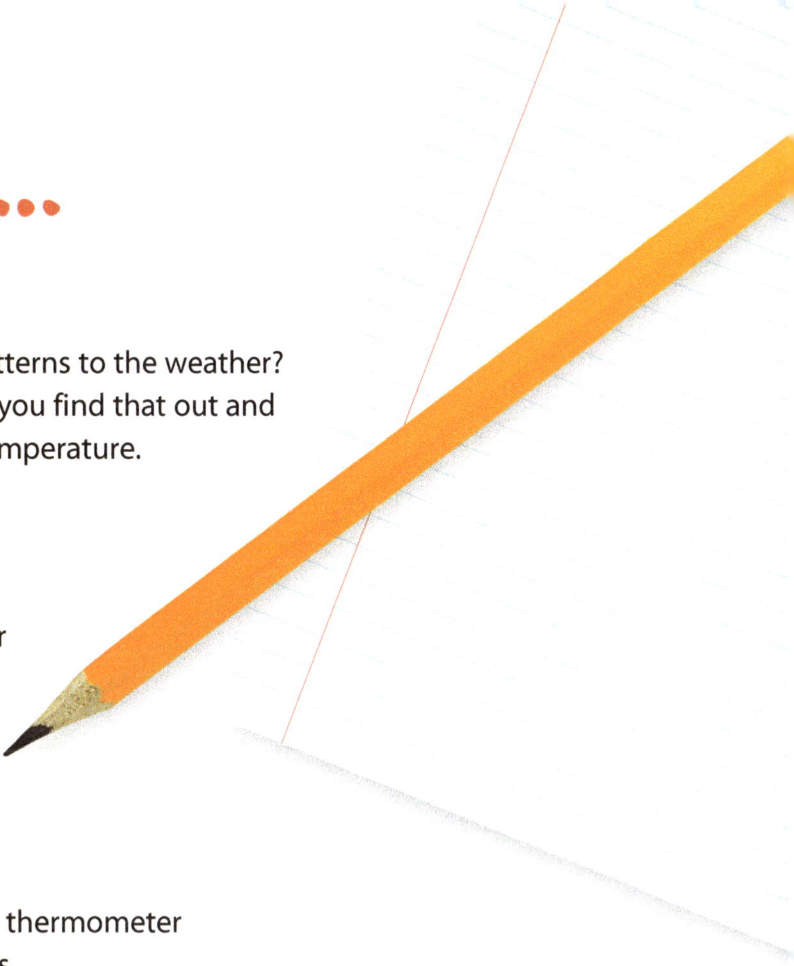

You Will Need:

- an outdoor thermometer
- paper
- a pencil

What To Do:

1. Ask an adult to put the thermometer in a safe place outdoors.
2. Check the temperature at five different times of day: early morning, midmorning, noon, afternoon, and evening. Write down the exact times and temperatures. When did the temperature rise and fall?
3. On another day, check and write down the temperature at the same five times of day.
4. Compare the temperatures for the two days. Did the temperatures rise and fall at the same times each day?

ANSWERS

From pages 16–17—Counting Triangles

They're all equilateral triangles.

From pages 18–19—Disappearing Squares

1. Here are four of the possible solutions.
   Take away the toothpicks shown in red.

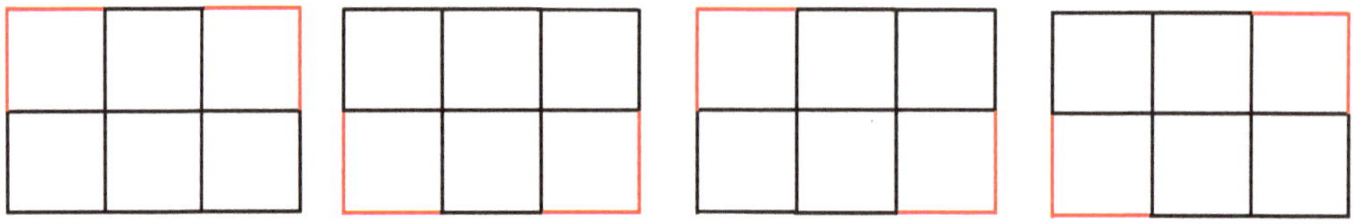

2. Here are two possible solutions.
   Take away the toothpicks shown in red.

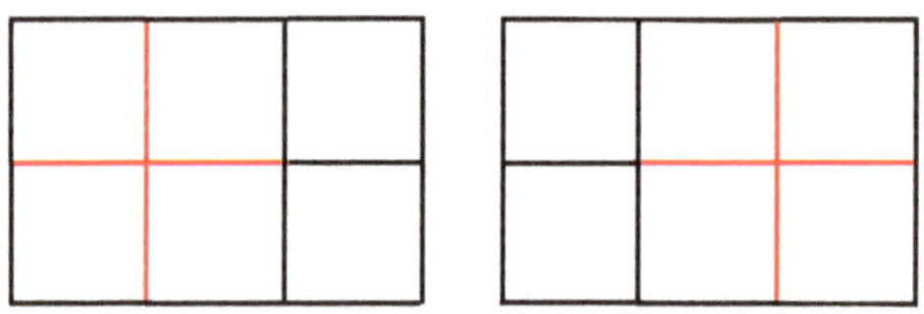

3. Move the red toothpicks as shown.

a.

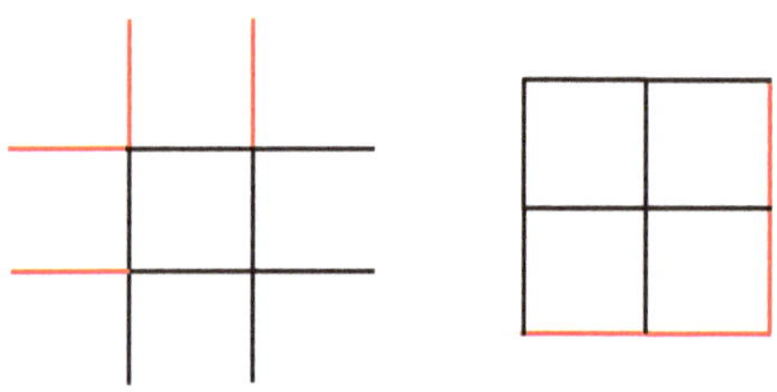

b.

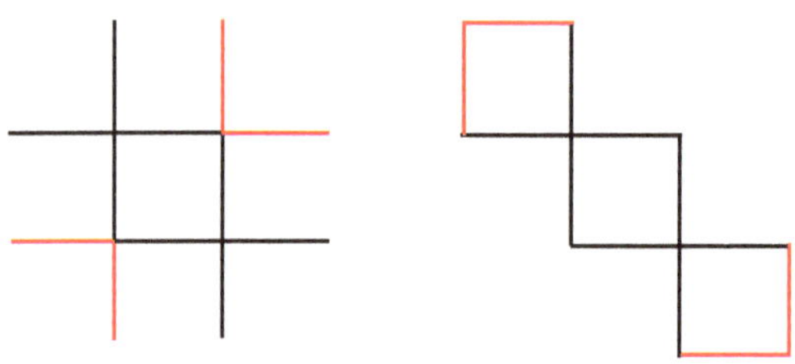

c.

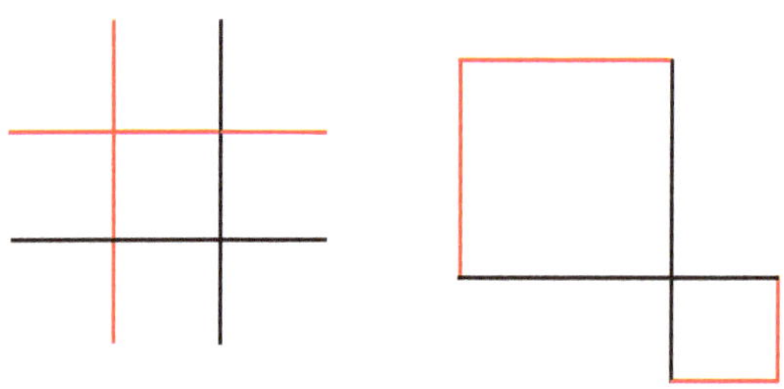

From pages 24–25—Polyominoes

Here is one way to make a 6 x 10 rectangle:

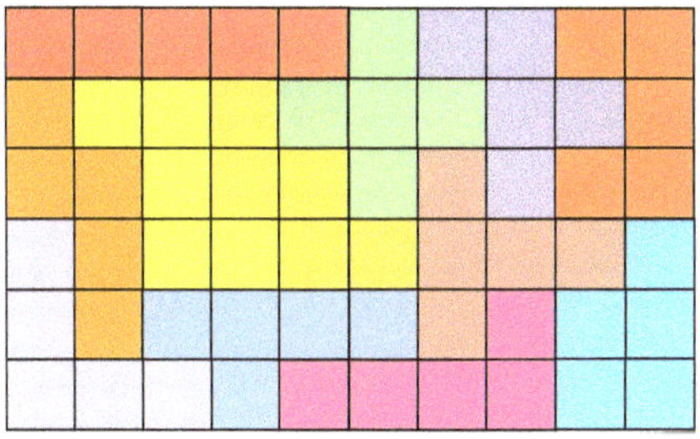

Here is one way to make a 5 x 12 rectangle:

From pages 32–35—Tangrams

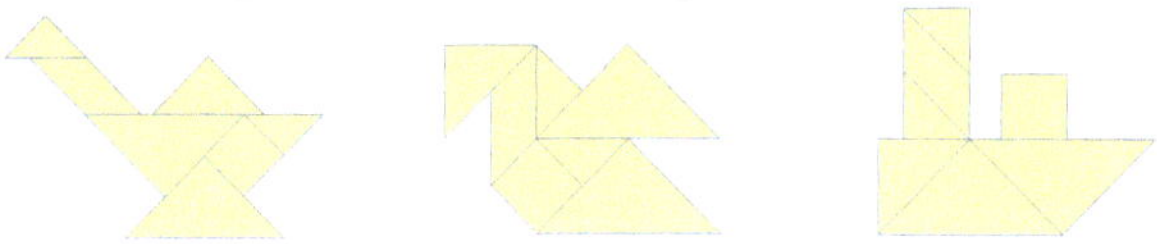

From pages 36–37—Polygons

All the shapes are symmetrical except the trapezoid and the parallelogram.

From page 43—Cube Puzzles

2 cubes, 8 sides:

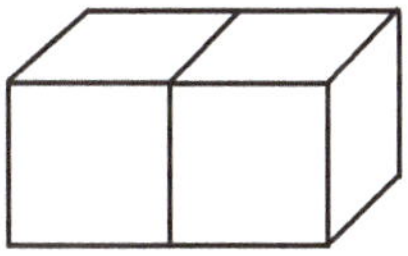

3 cubes, 11 sides:

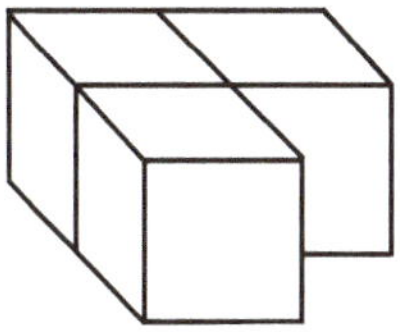

4 cubes, 12 sides:

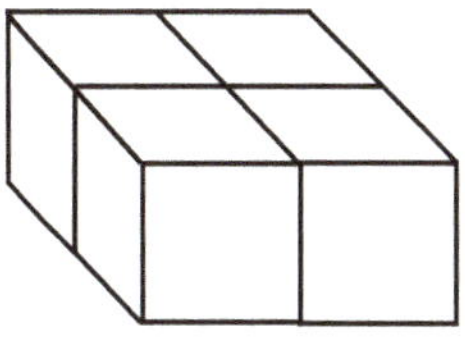

From pages 58–59—Count Like an Ancient Egyptian

These symbols in any order is a correct answer.

1. 

2. 

3. 

4. 

5. 

From pages 64–65—When in Rome…

1. II, VII, X

2. X, L, LXXX, CX, CXL

3. a. XVII

b. CLXIII

c. MDXXVIII

From pages 80–81—Pretty Patterns

purple, purple, yellow, red, red

From pages 84–85—The Art of Addition

Nine possible statements:

0 + 8 = 8	8 + 0 = 8
1 + 7 = 8	7 + 1 = 8
2 + 6 = 8	6 + 2 = 8
3 + 5 = 8	5 + 3 = 8
4 + 4 = 8	

From pages 86–87—Patterns in Addition

5 + 1 = 6
6 + 1 = 7
7 + 1 = 8
8 + 1 = 9
9 + 1 = 10

From pages 88–89—Addition Facts

1. When you add 0 to any number, you get the same number.
2. The answers to problems like 1 + 1, 2 + 2, and 3 + 3 form a diagonal line from the top left to the bottom right of the chart.

From pages 90–91—You Can Add!

1. 5 (4 + 1 = 5)
2. 9 (6 + 3 = 9)
3. 12 (7 + 5 = 12)
4. 20 (11 + 9 = 20)

From pages 94–95—More Magic Squares

Here are 10 other ways to get 34 by adding four numbers from the square:

1. The two middle numbers in the bottom row plus the two middle numbers in the top row:
   15 + 14 + 3 + 2 = 34
2. The two middle numbers in the first column plus the two middle numbers in the last column:
   5 + 9 + 8 + 12 = 34

3, 4, 5, 6. The four numbers in each quarter of the square:
16 + 3 + 5 + 10 = 34
2 + 13 + 11 + 8 = 34
9 + 6 + 4 + 15 = 34
7 + 12 + 14 + 1 = 34

7, 8, 9, 10. The four corner numbers of each 3 x 3 square within the magic square:
16 + 2 + 9 + 7 = 34
3 + 13 + 6 + 12 = 34
5 + 11 + 4 + 14 = 34
10 + 8 + 15 + 1 = 34

From pages 94–95—More Magic Squares: Try This...

3	14	13	0
8	5	6	11
4	9	10	7
15	2	1	12

5	0	7
6	4	2
1	8	3

From pages 100–101—You Can Subtract!

1. 3 (5 – 2 = 3)

2. 6 (9 – 3 = 6)

3. 8 (12 – 4 = 8)

4. 19 (20 – 1 = 19)

From pages 102–103—Backward Puzzles

Here is one possible solution for each puzzle:
Puzzle 2: 4 + 10 – 6 = 8
Puzzle 3: 6 + 4 – 7 = 3

From pages 104–105—The Minus Mystery

Gone to France. Working on book. Don't worry. Back soon.

From pages 112–115—Counting Squares

3 x 7 = 21

8 x 5 = 40

4 x 6 = 24

From pages 118–119—Square Numbers

You would need 49 candies, arranged in seven rows of seven, to make the next square number.

From pages 134–135—The Luck of the Dice

Usually 6, 7, and 8 come up most often;
4, 5, 9, and 10 come up next most often; and
2, 3, 11, and 12 come up least often.

From pages 148–149—Fun with Codes

Number code: "secret message" (Z = 1, Y = 2, etc.)
della Porta's code: "good luck"
"I like codes" in della Porta's code:

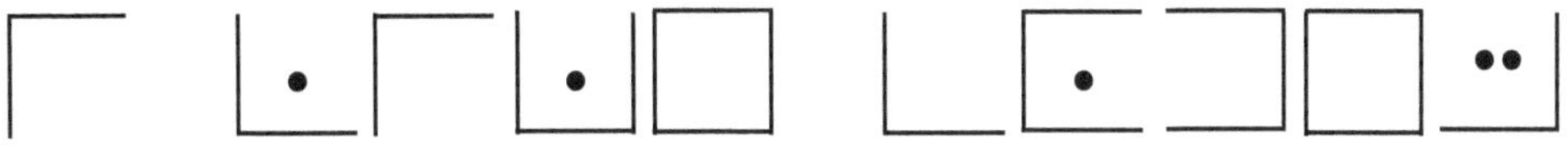

From pages 150–151—More Fun with Codes

1. the alphabet is turned around
2. it is all in how you look at it
3. decoding is fun
4. spaced differently and backward
5. the science museum

From pages 152–153—Statistics and Sports

1. Joel made the most goals in one game (4 goals in Game 1).
2. Fernando made the most total goals (8).
3. The basketball team scored 40 points.
4. Carmen got the most points (10).
5. There are 12 players on the team.

From pages 154–155—What Is a Graph?

Joel made 0 goals in Game 4.

Joel made the most goals in one game (4 goals in Game 1).

The blue section of the pie chart is the biggest.

That tells us that Carmen scored more points than any other player on the team.

From pages 168–169—What Is the Area?

The area of the castle is 46 squares. The area of the skyscraper is 38 squares. The area of the castle is bigger by 8 squares.

Glossary

Here are some of the words and phrases you read in this book. Many of them may be new to you. Some are hard to pronounce. But since you will see them again, they are good words to know. Next to each word, you will see how to say it correctly: **estimating** (EHS tuh MAYT ihng). The parts shown in capital letters are spoken more strongly than the rest of the word. Under each word are one or two sentences that tell what the word means.

A

abacus (AB uh kuhs)

An abacus is a counting tool. It has beads that slide on wires inside a wooden frame.

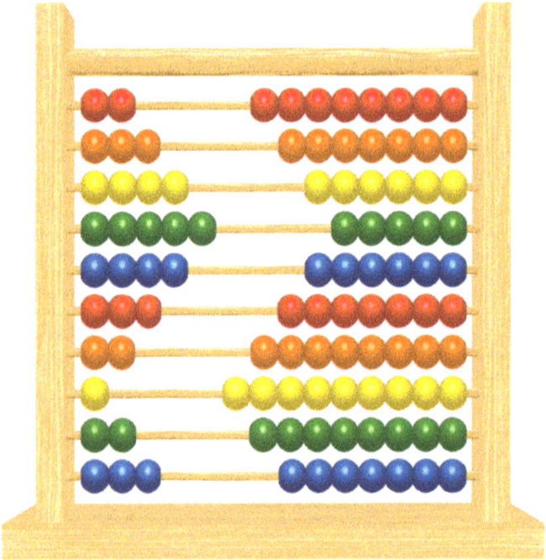

addition (uh DIHSH uhn)

Addition means putting numbers or groups of things together. An addition problem may look like this: $1 + 2 = 3$.

angle (ANG guhl)

An angle is the space between two lines that meet at a point.

area (AIR ee uh)

Area is the space inside a flat shape like a circle, square, or triangle.

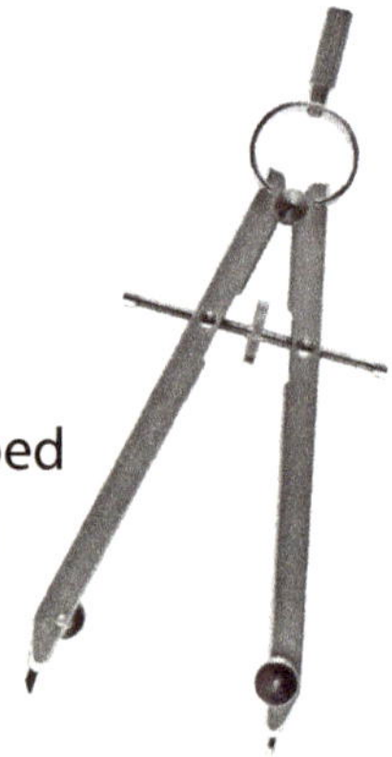

C

compass (KUHM puhs)

A compass is a V-shaped tool used for drawing circles.

cone (kohn)

A cone is a solid shape. One end is a circle, and the other end is pointed.

cube (kyoob)

A cube is a solid shape. It has six square sides, all equal.

cylinder (SIHL uhn duhr)

A cylinder is a solid shape. Its ends are two equal circles.

D

decagon (DEHK uh gahn)
A decagon is a flat shape with 10 sides and 10 angles.

division (duh VIHZH uhn)
Division is a way of breaking a number into smaller numbers. A division problem may look like this: $6 \div 2 = 3$.

E

estimating (EHS tuh MAYT ihng)
Estimating means using what you know to guess the size of something or the answer to a problem.

G

graph (graf)
A graph is a picture that shows relationships between numbers. Common kinds of graphs include bar graphs, line graphs, and pie charts.

H

heptagon (HEHP tuh gahn)
A heptagon is a flat shape with seven sides and seven angles.

hexagon (HEHK suh gahn)
A hexagon is a flat shape with six sides and six angles.

I

inch-pound system
(ihnch POWND SIHS tuhm)
The inch-pound system is a way of measuring things using units such as feet, pounds, and gallons instead of meters, grams, and liters.

infinite (IHN fuh niht)
Infinite means endless. There is an infinite supply of numbers.

L

length (lehngth)
Length is a distance in space. We measure length in a line, curved or straight.

M

measure (MEHZH uhr)
To measure is to find the size of something. You can measure many things, including length, weight, area, volume, time, and temperature.

metric system
(MEHT rihk SIHS tuhm)
The metric system is a way of measuring things based on 10's. Its units include the meter, liter, and gram.

multiplication
(MUHL tuh pluh KAY shuhn)
Multiplication is a fast way of adding numbers that are the same. A multiplication problem may look like this: 4 x 2 = 8.

N

nonagon (NAHN uh gahn)
A nonagon is a flat shape with nine sides and nine angles.

numerals (NOO muhr uhlz)
Numerals are written marks or symbols that stand for numbers.

O

octagon (AHK tuh gahn)
An octagon is a flat shape with eight sides and eight angles.

P

parallelogram
(PAR uh LEHL uh gram)
A parallelogram is a flat, four-sided shape. Its opposite sides are the same length and parallel.

pattern (PAT uhrn)
A pattern is a repeating arrangement of things or numbers.

pentagon (PEHN tuh gahn)
A pentagon is a flat shape with five sides and five angles.

place value (PLAYS VAL yoo)
Place value is the value, or meaning, of a numeral that comes from its position in a string of numerals. In the numeral 527, the 5 stands for five 100's. It has a higher place value than the 7, which stands for seven 1's. Whether a number means 1's, 10's, or more depends on the place it appears in a string of numerals.

polygon (PAHL ee gahn)
A polygon is a flat shape with three or more straight sides. Triangles, squares, and pentagons are all polygons.

Q

quadrilateral
(KWAHD ruh LAT uhr uhl)
A quadrilateral is a flat shape with four straight sides. Squares, rectangles, and parallelograms are all quadrilaterals.

S

sphere (sfihr)
A sphere is a perfectly round solid shape. Most balls are spheres.

squaring (SKWAIR ihng)
Squaring a number means multiplying it by itself.

statistics (stuh TIHS tihks)
Statistics are numbers that give information. They often describe groups of people or objects. Statistics about your family might include the number of people, their ages, their heights, and how many of them like chocolate ice cream.

subtraction (suhb TRAK shuhn)
Subtraction means finding the difference between two numbers or groups of things. A subtraction problem may look like this: $3 - 1 = 2$.

symmetrical (sih MEHT ruh kuhl)
Symmetrical means "balanced." A shape that has two matching halves is symmetrical.

T

trapezoid (TRAP uh zoyd)
A trapezoid is a flat, four-sided shape. Two of its sides are parallel, the same distance apart at all points. The other two sides are not parallel.

V

Venn diagram (VEHN DY uh gram)
A Venn diagram is a picture that groups objects in circles to show how they are alike (where the circles overlap) and how they are different (where the circles are separate).

volume (VAHL yuhm)
Volume is the amount of room inside a container or the amount a container will hold.

W

weight (wayt)
Weight is a measurement of how heavy something is.

Index

This index is an alphabetical list of important topics covered in this book. It will help you find information given in both words and pictures. To help you understand what an entry means, there is sometimes a helping word in parentheses, for example, **abacus** (calculator). If there is information in both words and pictures, you will see the words "with pictures" in parentheses after the page number.

Acknowledgments

Cover	© Kabachki.photo/Shutterstock; Miguel Angel Sánchez, Bright Agency
4-5	© MADDRAT/Shutterstock
6-7	© Peter Hermes Furian, Shutterstock
8-9	© Miles Boyer/Shutterstock
10-11	Estelle Carol; © Jukov studio/Shutterstock; © tam_odin/Shutterstock; © jalcaraz/Shutterstock; © Yganko/Shutterstock; © Best_photo_studio/ Shutterstock
12-13	CHILDCRAFT illustration; John Sandford and Steven Brayfield
14-15	Estelle Carol
16-17	CHILDCRAFT illustrations
18-19	George Suyeoka; © Winai Tepsuttinun/Shutterstock
20-23	George Suyeoka
24-25	Grahame Corbett; © Mega Pixel/Shutterstock; © GreenBelka/Shutterstock
26-27	© Sean Pavone/Shutterstock; © MichaelJayBerlin/ Shutterstock; © Gena96/Shutterstock; Steven Brayfield
28-29	© Nicholas Grey/Shutterstock
30-31	© Irina Nartova/Shutterstock; © Nikola Bilic, Shutterstock; Eileen Mueller Neill
32-33	© Kenishirotie/Shutterstock; © Fototocam/ Shutterstock; CHILDCRAFT illustrations; Rick Incrocci
34-35	CHILDCRAFT illustrations; Rick Incrocci
36-37	CHILDCRAFT illustrations
38-39	© Vitaly Zorkin/Shutterstock; © MNI/Shutterstock; CHILDCRAFT illustrations
40-41	CHILDCRAFT illustrations; © prostockphotos/ Shutterstock
42-43	Estelle Carol; Rebecca Schneider; © Voronin76/ Shutterstock; © photastic/Shutterstock; © 88studio/ Shutterstock
44-45	© Vasya Kobelev, Shutterstock
46-47	Larry Ross
48-49	Miguel Angel Sánchez, Bright Agency
50-51	© Ps_Studio21/Shutterstock; Bill and ludie Anderson
52-53	Bill and ludie Anderson
54-55	Jan Palmer; © Stephanie F Clarke, Shutterstock; © Dragon Images/Shutterstock
56-57	Bill and ludie Anderson
58-59	© WitthayaP/Shutterstock; © Ruslan Ivantsov, Shutterstock; John Sandford
60-61	Bill and Judie Anderson; CHILDCRAFT illustrations; Estelle Carol
62-63	© Pagina/Shutterstock; John Sandford; CHILDCRAFT illustrations
64-65	© Ruslan Ivantsov, Shutterstock; John Sandford
66-67	Tony Kenyon; Sue Snyder; © WESTOCK PRODUCTIONS/Shutterstock
68-69	CHILDCRAFT illustrations; Roberta Polfus
70-71	Roberta Polfus; Miguel Angel Sánchez, Bright Agency
72-73	© February_Love/Shutterstock
74-75	Estelle Carol
76-77	© New Africa/Shutterstock; Freddie Levin
78-79	© Dahlia/Shutterstock; Carl Whiting
80-81	Miguel Angel Sánchez, Bright Agency; © Nai_Pisage/ Shutterstock
82-83	© Ackab Photography/Shutterstock; © R. Mackay Photography/Shutterstock; Steven Brayfield
84-85	Roberta Polfus
86-87	Eileen Mueller Neill
88-89	© Vladimir Prusakov/Shutterstock; © Mcimage/ Shutterstock
90-91	Miguel Angel Sánchez, Bright Agency; © Urbanscape/Shutterstock; © NASA images/ Shutterstock; © A l l i b u m/Shutterstock; © pics five/ Shutterstock; © sangsiripech/Shutterstock
92-93	Janet Palmer; Malcolm Livingstone
94-95	Kinuko Craft
96-97	© Boltenkoff/Shutterstock
98-99	Roberta Polfus; © SP-Photo/Shutterstock
100-101	© donatas1205/Shutterstock; © Yellow Cat/ Shutterstock; Miguel Angel Sánchez, Bright Agency
102-105	Marleen Ekman
106-107	Malcolm Livingstone; © Hayati Kayhan/Shutterstock
108-109	© wk1003mike/Shutterstock; © MYMNY/Shutterstock
110-111	Roberta Polfus; © CrackerClips Stock Media/ Shutterstock
112-113	Kinuko Craft
114-115	Flexo Design; Kinuko Craft
116-117	Malcolm Livingstone
118-119	Steven Brayfield
120-121	Jenny Williams
122-123	Jenny Williams; © wk1003mike/Shutterstock
124-125	© Derya Cakirsoy, Shutterstock; © Ahmet Misirligul, Shutterstock; © wk1003mike/Shutterstock
126-127	Larrry Ross; Steven Brayfield
128-129	Miguel Angel Sánchez, Bright Agency; Roberta Polfus
130-131	© Iryna Usachova, Shutterstock; John Sandford
132-133	John Sandford
134-135	Malcolm Livingstone; Dennis Hockerman; © Leon Rafael/Shutterstock
136-137	Jenny Mumford; © orange vector/Shutterstock; © Sko Helen, Shutterstock
138-141	Dennis Hockerman
142-143	© jsalasberry/Shutterstock; Dennis Hockerman; Roberta Polfus
144-145	Dennis Hockerman; © pics five/Shutterstock; © KsanaGraphica/Shutterstock
146-147	Dennis Hockerman
148-149	Nigel Alexander
150-151	© Vasilyev Alexandr/Shutterstock; © Natykach Nataliia, Shutterstock; © silvae/Shutterstock
152-153	Roberta Polfus; © Lightspring/Shutterstock; © Fotokostic/Shutterstock
154-155	Roberta Polfus; CHILDCRAFT illustrations
156-157	© noicherrybeans/Shutterstock
158-159	Miguel Angel Sánchez, Bright Agency
160-161	The Louvre, Paris (CHILDCRAFT photo); Jack Wallen; © urfin/Shutterstock
162-163	Miguel Angel Sánchez, Bright Agency; © Ivelin Denev, Shutterstock; © Kenishirotie/Shutterstock
164-165	© supercavie/Shutterstock; © EHStockphoto/ Shutterstock; Roberta Polfus
166-167	© sagir/Shutterstock; © Runrun2/Shutterstock; Carl Whiting; John Sandford
168-169	© Ahmet Misirligul, Shutterstock; © WitthayaP/ Shutterstock; CHILDCRAFT illustrations
170-171	Miguel Angel Sánchez, Bright Agency; © zhekoss/ Shutterstock
172-173	© PixelSquid3d/Shutterstock; © Javier Brosch/ Shutterstock; © de2marco/Shutterstock; John Sandford; Eileen Mueller
174-175	John Sandford; Roberta Polfus
176-177	© Tomas Ragina/Shutterstock; © Boltenkoff/ Shutterstock; © Derya Cakirsoy, Shutterstock; Miguel Angel Sánchez, Bright Agency
178-179	Roberta Polfus
184-185	© Stephanie F Clarke, Shutterstock; © Gena96/ Shutterstock
188-189	© Peter Hermes Furian, Shutterstock; © Stephanie F Clarke, Shutterstock
190-191	© Peter Hermes Furian, Shutterstock

www.ingramcontent.com/pod-product-compliance
Lightning Source LLC
LaVergne TN
LVHW070524160926
848185LV00048B/1491
* 9 7 8 0 7 1 6 6 5 5 6 5 7 *